ELECTRICAL CONDUCTIVITY OF VITREOUS SUBSTANCES

ELEKTROPROVODNOST' STEKLOOBRAZNYKH VESHCHESTV

ЭЛЕКТРОПРОВОДНОСТЬ СТЕКЛООБРАЗНЫХ ВЕЩЕСТВ

ELECTRICAL CONDUCTIVITY OF VITREOUS SUBSTANCES

Rudol'f L. Myuller

Leningrad State University
Leningrad, USSR

Translated from Russian by
S. Drake and C. F. Drake

 Springer Science+Business Media, LLC 1971

The original Russian text, a collection of articles by the late Professor
R. L. Myuller selected by an editorial committee consisting of Z. U.
Borisova, O. V. Mazurin, and V. S. Molchanov, was published by Leningrad
University Press in 1968 for the A. A. Zhdanov Leningrad State University.
The English translation is published under an agreement with Mezhdunarod-
naya Kniga, the Soviet book export agency.

Library of Congress Catalog Card Number 74-128508

ISBN 978-1-4757-5064-5   ISBN 978-1-4757-5062-1 (eBook)
DOI 10.1007/978-1-4757-5062-1

© 1971 Springer Science+Business Media New York
Originally published by Consultants Bureau, New York in 1971.

Preface

Professor R. L. Myuller has, during his many years of study of the chemical nature of glass, made a fundamental contribution to the development of glass science. Thus, from his analysis of the results of conductivity studies of simple and complex oxide glasses he predicted the chemical inhomogeneity of complex glasses, which has been proved 20-25 years later by x-ray and electron-microscope studies.

Myuller's studies of the conductivity and chemical resistance of glass are particularly important. He devoted much attention to the study of semiconducting glasses and developed an original concept of the conductivity of semiconductors with a low current-carrier mobility.

The author himself intended to produce a unified summary of his work in a monograph but his tragic death intervened.

From the rich scientific legacy he left (in electrochemistry, thermochemistry, viscosity, solution kinetics of glass, etc.), we have selected for publication here his work on the conductivity and structure of glass. A bibliography at the end of the book lists all Myuller's published works.

We have drawn very little from the articles on the conductivity of glasses published in the earliest period (one article, "The electrical conductivity of glasses," summarizes a large amount of experimental material); nor have we reproduced in the present selection the work published in Solid State Chemistry* and Vestnik, Leningrad State University, No. 22, 1962.

The author's articles are characterized by a close interrelation; each work is based on the results and conclusions of several previous works. This is particularly typical of the series on the conductivity of solid ionic-covalent materials which comprises eleven articles. This series has been published previously in various journals and is given here in its entirety. We have retained the author's numbering (I to XI) in this case. We have also kept his symbols for various parameters although there are differences in the designation of the same parameter in different articles of different periods. Thus, the electrical conductivity in the early period is shown as $\varkappa$, but in the later articles as σ.

In preparing the manuscript for publication we have corrected only the errors and inaccuracies which had passed unnoticed in previous editions. Otherwise, the author's text is unchanged. Essential explanations and additional information are presented in footnotes. In the lists of cited papers those which appear in the present selection are distinguished by a black circle (●).

Z. U. Borisova

O. V. Mazurin

V. S. Molchanov

---

*R. Yu. Muller, Solid State Chemistry (Z. Yu. Borisova, ed.) Leningr. Gos. Univ. Izd, Leningr. (1965) [Eng. trans., Consultants Bureau, N.Y. (1966)].

Contents

PART 1

ELECTRICAL CONDUCTIVITY OF IONIC GLASSES

Electrical Conductivity of Glasses*

1. A study of the physical properties of matter was made somewhat simpler by the establishment of a system of classification according to the state of aggregation; gas, liquid, or solid. Such mechanical properties as the immutability of form, constancy of volume, and ability to flow, can be used as criteria of these subdivisions and the state of aggregation can then be attributed on the basis of the presence or absence of these properties. It is essential to these considerations that the system in question be in a state of thermodynamic equilibrium and it is then possible to evoke general phenomenological principles characteristic of each of these three states of aggregation. Having ennumerated all the stable equilibrium states of matter there remain to be examined the states of aggregation which represent nonequilibrium systems. In particular, the vitreous state does not correspond to any of these states of aggregation and is an effectively distinct, "fourth" state of aggregation.

In addition to this classical classification of states, another method of subdivisions is very important in chemistry. In order to solve problems relating to the atomic and molecular structure of matter, knowledge of the degree of ordering in the distribution of the particles is often quite essential. [1]. In this case it is easy to indicate the two extreme cases:

A. The case where there is complete statistical disorder characteristic of the ideal gas. The position in space of any particle is not associated with the orientation or position of the neighboring particles.

B. The case where there is an ordered state corresponding to an ideal crystal. The position of each particle is now wholly determined by the orientation and position of the neighboring particles.

We do not, in fact, ever encounter these extreme states but we do have the opportunity of studying substances in states very close to the extremes. For the most part, however, we are dealing with an intermediate or relatively ordered distribution of particles. From this point of view the study of the transition between order and disorder in the distribution of particles becomes extremely important. Considerable attention has in recent times been directed to the transition of a substance from the ordered solid crystalline state to the disordered liquid state. It is known that an increase in temperature makes the force fields become weaker owing to some uncoupling of the particles. Thus, a decrease in the spatial density, an increase in particle vibration, and a weakening of the orienting forces, destroys the order in the distribution of the particles.

By changing the pressure and temperature we can trace a continuous transition from the relatively oriented arrangement of particles in the liquid state to the disordered distribution of particles in the gaseous state. However, we do not have at present any basis for assuming that a continuous change is possible from the ordered crystalline state to the disordered liquid or

*R. L. Myuller, Uch. Zap. Leningr. Gos. Univ. No. 54 (5), 159 (1940).

gaseous state by changing only the temperature and pressure. From this point of view it is of considerable interest to ask whether a continuous transition is feasible by a continuous change of some other important factor such as the concentration.

In the case of a single-component system such a problem does not, of course, arise. However, when there are two components the possibility of a continuous transition from an ordered crystalline state to a disordered state analogous to that of a liquid can be considered. In fact we can select two such components, one of which will have particles which quite easily become oriented relative to each other on solidification and as a result easily produce a crystalline lattice, while the other will be immeasurably more difficult to crystallize or may even not be in a crystalline state when the melt solidifies. After solidification of such a two-component system we would then have in the medium a disordered arrangement of particles of one type in contrast with an ordered distribution of the readily orientatable particles of the second type. Of course, with an increase in the concentration of the latter type, orientation and distributional order must increase as a consequence of the increased interaction of these particles. With a sufficiently low concentration of the poorly crystallizing component the onset of normal crystallization of the second component would be expected.

One example of such a two-component system is a melt containing boric anhydride and borates of the alkali metals. Boric anhydride can be obtained in a crystalline state from boric acid [2]. Crystals cannot be obtained directly by freezing a boric anhydride melt. Thus, boric anhydride is clearly a component the particles of which are not readily oriented in a lattice when the melt is frozen. On the other hand, borates of the alkali metals, for example $NaBO_2$, $Na_2B_4O_7$, $Na_2B_8O_{13}$, $Li_2B_4O_7$, etc., are easily crystallized salts whose particles are readily oriented.

It is easy to obtain a similar system from a melt of boric anhydride with an alkali metal oxide. The mixture so obtained is of a complex character. It contains various boric anhydride polymers and various alkali borates. It is certain, however, that the system contains, in addition to the poorly oriented particles of the polymerized boric anhydride, easily mutually orientatable particles of salt-like borates in an amount corresponding to the alkali oxide content.

This system is analogous to others which includes the so-called glass-forming oxides, SiO_2, P_2O_5, etc. A very important factor in obtaining such systems in the vitreous state is the cooling rate in the vicinity of the softening point where prolonged complex reorientations and very slow displacements of the particles occur. In general the state obtained here is not equilibrated. Despite this, as well as the considerable practical value, such unique semi-amorphous formations can be very interesting objects for studying intermediate states of a substance during its transition from a disordered to an ordered state and vice versa. In particular, such systems have already played a well known role, and will undoubtedly contribute even more to an explanation of the mechanism of crystallization. As regards the practical value of systematically directed studies, the important advances in the field of glass technology from the time of the first work of Winkelmann and Schott have always been the result of such systematic physicochemical studies [3]. It is essential in this case to make such a choice of the parameters to be measured that their dependence on composition may be used to resolve the problem in question.

2. The chemical literature is exceptionally rich in work devoted to a study of the dependence of properties on composition. It is sufficient to note here the extensive studies of the Kurnakov school, the systematic studies on vitreous systems carried out by the Grebenshchikov school, and the glass studies of Tamman, and Turner et al. The most diverse parameters were studied in these investigations. Recently, x-ray analysis has become vital in the study of structure.

We used the electrical conductivity as the parameter to be studied in its dependence on composition. It is well known that the electrical conductivity of glass has an ionic character and is caused by the presence of a small proportion of the ions in free translational motion. Thus, sodium ions move in normal glass. The idea of organizing a systematic study of the electrical conductivity of a glass of the simplest composition belongs to Shchukarev who propounded the concept that the study of glasses, treated as super-cooled solutions, would extend our knowledge of the electrochemistry of solutions. As a result of our work this concept later yielded an interesting and fundamental result: the establishing of a mobility series for unsolvated ions in solid vitreous media. This series was found to be in the reverse order to that known in liquid aqueous solutions. Later we discovered that Frenkel's theory of the electrical conductivity of crystals was applicable to alkali-rich glasses. According to this theory, in a solid at any given moment there is an equilibrium between the ions which are firmly bonded at the lattice points and the ions which are detached from the lattice points as a result of energy fluctuations and are in translational motion having strayed to interstitial positions. It is clear that the specific electrical conductivity will be proportional to the concentration of such freely moving ions in the intralattice space. The temperature coefficient of the electrical conductivity will be directly associated with the energy required to detach the ion from the lattice points.

Thus, if the ratio of the concentration of mobile sodium ions $[\overrightarrow{Na^+}]$ to the concentration of sodium ions at the lattice points $[Na^+]$ is taken, from Frenkel's theory, to be

$$\frac{[\overrightarrow{Na^+}]}{[Na^+]} = e^{-\frac{E}{2RT}},$$

where E is the energy of the displacement of the ion from the lattice point; R is the gas constant; T is the absolute temperature, then for the specific electrical conductivity we obtain, as a first approximation, the following expression

$$\varkappa \sim [\overrightarrow{Na^+}] = [Na^+] \cdot e^{-\frac{E}{2RT}}, \tag{1}$$

where $\varkappa$ is the specific electrical conductivity.

By studying the dependence of specific electrical conductivity on temperature we can determine the energy E, which represents the binding strength of alkali metal ions to the lattice points of the complex and peculiarly distorted glass lattice. At present there is evidence to support the idea that in solid glass we have some irregularity in the arrangement of the potential energy minima in which particles are located, analogous to that which occurs in crystals. The strength with which the ions are held in these potential minima can also be determined from the experimental measurements of the temperature dependence of the electrical conductivity. The bonding strength of the sodium ions in a glass depends on its structure, and, in particular, is determined by the relative arrangement of the more easily orientatable ions of the salt-like component of the glass. By studying the electrical conductivity of different specimens of glass with a gradually changing composition we can follow the course of the change in the energy of displacement of ions from the lattice points.

It is doubtful whether the alkali metal ion is strongly bonded to the nonpolar boric anhydride molecules. Moreover, the strong polarizing action of the nucleus of the boron atom on the oxygen atoms eliminates any significant polarizing action of the alkali ions on the same oxygen atoms. It can be said that the sodium ions are bonded by the mainly electrostatic forces of the anions. By changing the concentration of the alkali oxide we change the relative concentration of easily orientatable alkali borate ions in the boric anhydride matrix which is difficult to crystallize. We thus arrive at the problem of the distribution of the particles.

These considerations make it clear why the study of conductivity in vitreous boric oxide—alkali systems is of interest, both from the point of view of studying the problem of the transition of matter from a disordered to an ordered state, and for understanding the nature of the vitreous state itself.

In addition to its serious theoretical interest, the systematic study of the electrical conductivity of glass is of great practical value in electrical technology. There is no opportunity here to consider in more detail the technical problems and we shall only indicate the practical value of studying this field of electrochemistry in order to determine the optimum composition for glass used in the manufacture of cathode-ray tubes. The glass in these is altered as a result of dc electrolysis. The understanding of electrical conductivity has no less a practical value in determining the composition of low-melting glasses which may need to be either good or, on the other hand, bad electrical conductors. An example of the first case is the practical problem of standardizing the composition of glass electrodes which are, of course, extremely valuable technically in automating control and monitoring of production processes. We meet the second case in the problem of producing insulating enamels for insulation.

3. The first work on the measurement of electrical conductivity in glass was carried out in the middle of the last century [6]. In 1884 Warburg's classical work establishing the ionic nature of conduction in glass was published [7]. Both in this and in all subsequent work the authors directed their investigations to uncovering the physical principles of the electrical conductivity of glasses, but were at the same time completely uninterested in their chemical composition and only concerned that the glasses were sufficiently rich in alkalis to provide electrical conductivity. Technical glasses of quite complex compositions were exclusively studied and hence the effect on the conductivity of the type and concentration of the individual components was left completely unexplained. The valuable, if somewhat unique, work of Ambronn [8] and Gehlhoff and Thomas [9] did not solve these problems. Ambronn studied the relatively complex three-component sodium calcium silicate glasses and as a result was not able to reveal the effect of each component on the conductivity. Gehlhoff and Thomas limited their intentions to narrow technical aims and in solving the problem they utilized a series of methodological simplifications. Serious doubt has been expressed concerning the validity of Gehlhoff and Thomas's list of glass components arranged in ascending order of contribution to the electrical conductivity.

In 1926 we began a systematic study of the dependence of electrical conductivity on composition of a series of solid glasses. The compositions were made progressively more complex and more stringent methods for measuring the conductivity were used.

The method we developed for measuring the conductivity involved the use of the active sodium amalgam electrodes, recommended by Warburg to eliminate electrochemical polarization. We included the method of measuring glass platelets described by LeBlanc and Kirschbaum in [10], and applied Curie's electrometric method for the measurement of conductivity [11]. We introduced a number of improvements, but neither these nor the other experimental details can be considered in detail here since they have been described fully elsewhere [4,12].

In developing our method we had seriously to consider the reproducibility of our results which were checked, not only by repeating the measurements at various temperatures for each specimen of glass but also by studying different specimens of glass of the same composition and by using three different methods for the electrometric determination of the conductivity of the glass under consideration. The experimental study showed that this method enabled us to determine the absolute values of the conductivity of glasses of different composition with a mean error of 10-15%. Realizing that the range of the measured values of conductivity extended from 10^{-17} to 10^{-5} $\Omega^{-1} \cdot cm^{-1}$, i.e. that the specific electrical conductivity changed by twelve orders

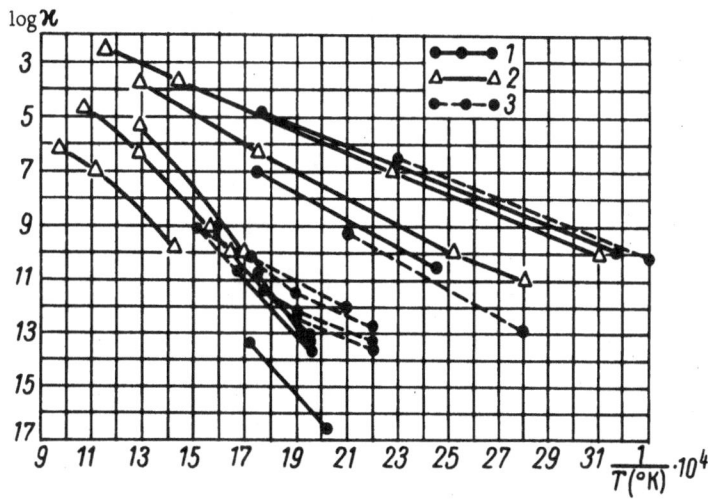

Fig. 1

of magnitude, the reproducibility we obtained must be considered entirely satisfactory. The good reproducibility of our results is supported by the work of Thomas [13] (school of Parks) who, a year after the publication of our work on the system $Na_2O-B_2O_3$, published his work on the electrical conductivity of the same system. This work was quite independent of ours and made use of a completely different method. Thomas' results (Fig. 1, curve 2), often made at higher temperatures, are in good agreement with ours (curve 1). Finally, the work of Spaght and Clark [14] which appeared three years later and was concerned with the same system, $M_2O-B_2O_3$, also confirmed our data (curve 3). It is true that in the case of boric anhydride these authors obtained rather strange results which disagreed, not only with our data, but also with that of Thomas, their colleague. We showed later, on the basis of our completed measurements, that this discrepancy in the data for boric anhydride was clearly the result of an error by Spaght and Clark in their measurement of the primary current, in that they did not establish steady conductivity conditions.

As a result of a critical analysis of our results by contrasting and comparing them with the results of other authors, we were able to conclude that we had a satisfactory method at our disposal.

It was clear that the experimental conductivity data we had obtained could be used with confidence as a reliable basis for solving the problems we had posed.

4. Starting with the simplest two-component vitreous systems, $M_2O-B_2O_3$ and $MO-B_2O_3$, containing univalent alkali metal ions and divalent barium and zinc ions, we undertook systematic studies of their conductivity. More recently we have measured the conductivity of complex three-component systems of the type $M_2'O-M_2''O-B_2O_3$. We are at present studying the conductivity of the system $Ag_2O-B_2O_3$.

As a result of many years of work in which Markin, Brodskaya, Tatarinova, Taking, and Shchegoleva participated, there are now more than 3500 measurements of the electrical conductivity of boron glasses and the following vitreous two-component systems have been investigated: $Li_2O-B_2O_3$, $Na_2O-B_2O_3$, $K_2O-B_2O_3$, $Rb_2O-B_2O_3$, $Cs_2O-B_2O_3$, $BaO-B_2O_3$, $ZnO-B_2O_3$; and the three-component systems: $Li_2O-Na_2O-B_2O_3$, $Na_2O-K_2O-B_2O_3$, $Li_2O-K_2O-B_2O_3$, and $Na_2O-BaO-B_2O_3$.

The experimental problems associated with the production of glass specimens of specified composition and with their analysis and treatment have been discussed in particular papers [4] and are not considered here.

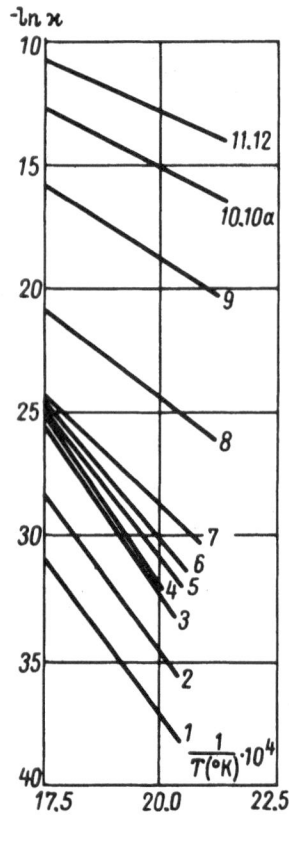

Fig. 2

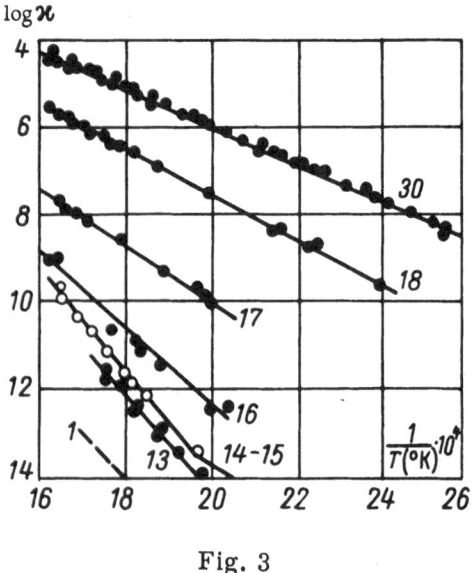

Fig. 3

5. The two-component systems referred to above exhibited, as expected, over a wide range of temperatures a linear dependence of the log specific electrical conductivity ln $\varkappa$ on the reciprocal of the absolute temperature 1/T:

$$\ln \varkappa = -\frac{E}{2RT} + B. \tag{2}$$

This expression is completely consistent with the theoretical relationship of Eq. (1) above and may be obtained from the latter by simply taking logarithms. It is clear that, by plotting log specific electrical conductivity against the reciprocal of the absolute temperature, we can easily determine from the slope of the straight lines so obtained the value of the very important value E, the energy of ion displacement. The graphs for the systems $Na_2O-B_2O_3$ (Fig. 2), $Li_2O-B_2O_3$ (Fig. 3), and $K_2O-B_2O_3$ (Fig. 4) [4] shown here illustrate this linear dependence.

The last two figures provide an opportunity to verify the good agreement between the experimental points and the theoretical straight lines. All the graphs shown here contain straight lines which are numbered in ascending order as the alkali metal ion concentration in the specimens increases. We now see that with an increase in the ion content, i.e. in the salt-like part of the glasses, the electrical conductivity increases rapidly and the slope of the lines decreases, i.e. the displacement energy E falls. The calculated values of this energy call for some attention. With a low concentration of the salt-like component the displacement energy is high and approaches 100 kcal, i.e. approaches the dissociation energy of the simplest salt-like molecules (containing univalent ions). With a high content of the salt-like component the ion displacement energy drops to 40 kcal, which is, in fact, identical with the ion displacement energy in ionic crystalline lattices. Thus a direct knowledge of the values of ion displacement energies indicates

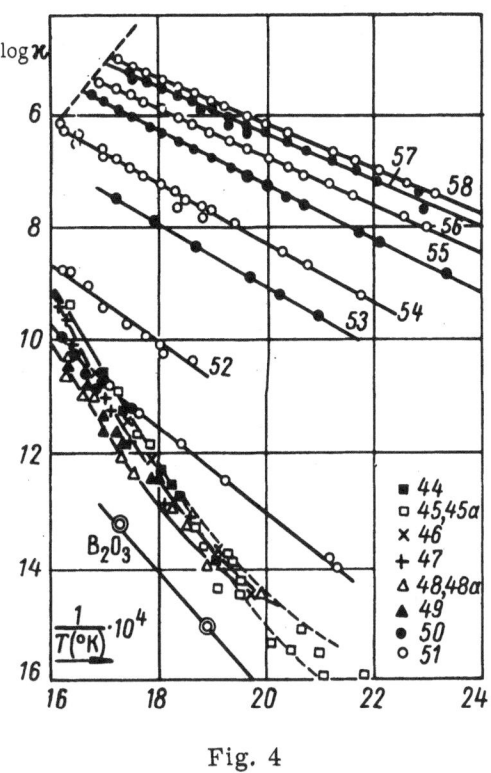

Fig. 4

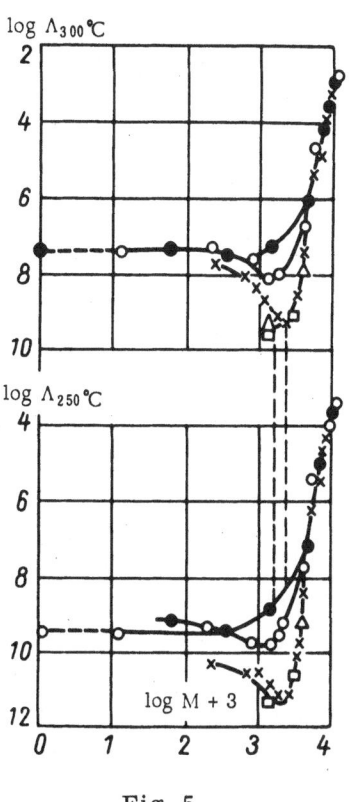

Fig. 5

that in glass low in alkali the salt-like component exists in a state close to that of a molecular dispersion. With a high concentration of the salt-like component we have the structure of this component approximating to that of a crystal. Ion motion clearly occurs in the first case within the boric anhydride matrix and in the second in the interstices of an ionic lattice which is, of course, slightly distorted in glasses.

It follows that the number of freely moving ions in the intralattice space will be significantly greater than in the boric anhydride matrix. In fact, as the experiments show, an energy expenditure of 40 kcal is required in the first case, while in the second it is considerably higher. This makes sense if we consider that in the absence of any significant reaction between the ions and the boric anhydride (see above), a considerable energy must be expended in dissociation which involves the removal of the cation to a very great distance from the anion in the context of the boric anhydride matrix. The energy of dissociation of an ion in the lattice is considerably less since the ion displacement is facilitated by the electrostatic field of the neighboring ions. As a result of the quite considerable decrease in the dissociation energy with an increase in the alkali content of a glass and hence of a considerable increase in the number of freely moving ions, a significant rise in the electrical conductivity of the glass would be expected.

The lower energy in this case is the result of the electrostatic interaction of the ions. Therefore we expect to see the beginning of the rise in electrical conductivity when there is a sufficiently close approach of the salt-like molecules to each other.

Experimentally it is found that the conductivity increases sharply with the alkali concentration and that the beginning of the increase corresponds to an average approach distance of the salt-like molecules to each other approximating to the diameter of boric anhydride molecules.

The most convenient parameter for expressing the concentration dependence of electrical conductivity, of course, is molar electrical conductivity

$$\Lambda = \ln \frac{\varkappa}{m} = - \frac{E}{2RT} + (B - \ln m), \qquad (3)$$

where m is the borate concentration expressed in mole/cm^3. The log molar electrical conductivity is plotted in Fig. 5 against the log of the borate concentration M, expressed in mole/liter of glass.

It is clear from Fig. 5 that the molar electrical conductivity changes slowly at first and then increases sharply on reaching some critical concentration. With an increase in the alkali concentration by 15-20 times, the molar conductivity of potassium borate glasses at 250°C increases by a factor of 10^8. We thus observe an extremely rapid rise in conductivity as the salt-like molecules approach closer to each other.

If we calculate the average distance between the salt-like molecules at the critical concentration from the analytical results and from the density of the glass, we find that at this critical concentration the salt-like molecules are at a distance from each other which is approximately the measured diameter of the boric anhydride molecules [15].

6. We have thus established experimentally and theoretically that an abrupt increase by more than 10^6 in the conductivity occurs at a certain critical concentration corresponding to the limiting approach of the salt-like molecules in the glass. It is an extremely important research result of relevance to the technological electrochemistry of glass. It is certainly of practical interest that there is a specific critical concentration of the alkali oxide (in sodium glasses it is about 5% of sodium oxide) above which an insignificant change in the alkali concentration in the glass produces an exceptionally large change in the conductivity. Of course, it is quite essential to know the principles which determine the quantitative dependence of conductivity on concentration.

Since the conductivity of glass is basically determined by the energy of ion displacement, the establishing of the principle which determines the dependence of molar electrial conductivity on ionic concentration must at the same time lead to the discovery of the quantitative dependence of the ion-displacement energy on the same concentration. This connection between the ion-displacement energy and concentration must, above all, characterize the type and magnitude of the electrostatic interaction of nearby ions during a continuous change in the concentration of these ions. Moreover, it provides an interesting explanation of the presence or absence of the effect of individual features peculiar to various alkali ions and the role which is played by the boric anhydride. It is true that the function of the anhydride is fairly complex, since it is a component part of the complex anion and at the same time is the medium.

The work referred to above emphasizes the undoubtedly significant practical and theoretical interest in the connection between ion-displacement energy and ion concentration in glasses.

An analysis of the experimental data on the electrical conductivity of borate glasses indicates that they satisfy the empirical relationship [5]

$$\ln \Lambda = C_1 - \frac{C_2}{T m^{\frac{1}{x}}}, \qquad (4)$$

where Λ is the molar electrical conductivity; C_1 is a constant which depends on the nature of the alkali oxide and lies within the narrow limits $8 \leq C_1 \leq 11$; and C_2 and x are constants which depend on the nature of the glass-forming oxide. All these relationships can be reproduced quantitatively in the form shown in Table 1.

TABLE 1

Boric anhydride	C_1	C_2	x
B_2O_3 SiO_2	8—11	900 2400	2 4

Table 1 also shows the values of the parameters C_1, C_2, and x for silicate glasses, obtained by recalculating the experimental data of Turner and his colleagues [16]. Not only the general principle but also the character of the rise in electrical conductivity with an increase in concentration in the system Na_2O-SiO_2 studied by Turner, indicates the generality of the nature of electrical conductivity of borate and silicate glasses. Above all, this can be verified by glancing at Fig. 6. This represents the logarithmic dependence of the molar electrical conductivity on molar concentration of sodium silicate in the vitreous system $SiO_2-Na_2SiO_3$, and also of borax in the vitreous system $B_4O_6-Na_2B_4O_7$. There is a lack of data in the silicate system for compositions which are intermediate between fused quartz and glasses rich in sodium. This is explained by difficulties in obtaining such glasses in a sufficiently homogeneous form. In spite of this we may assume a similar type of dependence of electrical conductivity on concentration in borate and silicate glasses.

From Eqs (3) and (4) we obtain a relationship between displacement energy and concentration:

$$E = \frac{\text{const}}{\sqrt[x]{m}}. \tag{5}$$

The value of the constant in the case of borate glasses is approximately 3600 and in the case of silicate glasses is about 9600. The last expression is quite empirical and while on going from the borate to the silicate system x changes from a value of approximately 2 to 4, there is clearly also a change in the constant. This bears out the complex character of the principle but at present it would be premature to look for the physical chemical significance. No doubt, with the extension of the study of the chemical bond in condensed systems this empirical principle will in time be explained. It is possible that it may serve to lead to an explanation of the complex and interrelated nature of the interaction forces between particles (in particular ions) when they are in close proximity.

7. The investigation of the electrical conductivity of two-component vitreous systems provides new experimental material which leads to a deeper knowledge of the electrochemistry of glasses containing a salt-like component. The study of the conductivity of three-component vitreous systems containing two alkali oxides is of considerably greater practical importance. Research in this direction is interesting, too, in clarifying the nature of the vitreous state of such systems.

We measured the conductivity of the systems $Li_2O-Na_2O-B_2O_3$, $Li_2O-K_2O-B_2O_3$, as well as the technically most interesting system $Na_2O-K_2O-B_2O_3$. As in the two-component system, the log of the specific electrical conductivity retains its linear dependence on the reciprocal temperature over a wide temperature range.

The fundamental result of this series of studies is the clarification of the dependence of the conductivity on the composition of the glasses.

It has been established that, during the gradual equimolecular substitution in the glass of one form of alkali ion by another form, no simple additive dependence of the electrical conductivity of the glass is observed (with a linear character throughout the change in relative content

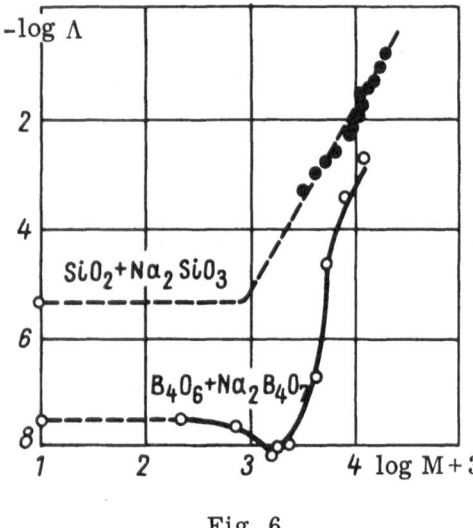

Fig. 6

of the alkali components). The additive dependence was found to be more complex. Figure 8 in the paper by Brodskaya and Tatarinova [19] serves to illustrate such a dependence. It shows the curve (continuous line) of the concentration dependence of the log specific electrical conductivity at 300°C for the vitreous system $Na_2O-K_2O-B_2O_3$ with a molar fraction of the total alkali oxides of 0.75.

The percentage molar content of sodium oxide in the glass plotted along the abscissa is expressed in the following units:

$$\frac{\text{moles } Na_2O}{\text{moles} Na_2O + \text{moles } K_2O} \cdot 100.$$

The curve has a minimum, and both of its decremental sections correspond to the sharp drop in the electrical conductivity with a decrease in the concentration of each of the alkali ions. This is easy to verify if one compares the experimental curve in question with the calculated curve (Fig. 8, black points) which corresponds to the log of the total specific electrical conductivity of two-component glasses each of which contains one alkali oxide at the same concentration as in the three-component system in question. We are dealing with the case of a glass with a high ion content. Therefore, the ion movement occurs in the way discussed above for this type of glass, that is in the interstitial positions in the ionic lattice. Since the total, sodium plus potassium, ion content does not change it would be natural to expect that their electrostatic interaction in a uniform mixture of both ions would produce an unchanged ease of movement for both types of ions, independent of their relative concentration. The experiment shows exactly the reverse and indicates that the functional dependence of electrical conductivity on composition in the two-component systems is retained in the three-component system. Allowing for the existence of a connection between electrical conductivity and structure in glasses we may conclude that the structural peculiarities of the two-component systems is retained in the three-component system. This is possible only in the case where such a system is nonuniform and is a heterogeneous mixture of the corresponding two-component systems.

If this conclusion is true it follows that the displacement energy of the excess ions in the three-component system must be equal to the displacement energy of the corresponding ions in the two-component systems at an equal volume concentration. The experiment confirms this conclusion which is easily verified if we consider Fig. 7 which shows the concentration dependence of the displacement energy of the potassium and lithium ions [M] = [K] + [Li] in the three-component system $Li_2O-K_2O-B_2O_3$ (continuous line) and the displacement energy of lithium and potassium ions at the same concentration in the appropriate two-component systems (broken lines). The agreement can be considered satisfactory.

The deviations observed here are rather interesting. Without going into detail it can nevertheless be seen that they are probably to be explained as follows. Consider the presence of a different content of ions of the two types. A change will be inevitable in the physical conditions under which ion displacement occurs, consequent on the substitution of the excess boric anhydride by the second salt-like component on the transition from the two- to three-component system.

8. As we have already noted above, three-component systems are the most interesting from a practical point of view. They exhibit, in the simplest form, the interaction between the two salt-like components which is observed in technical glasses.

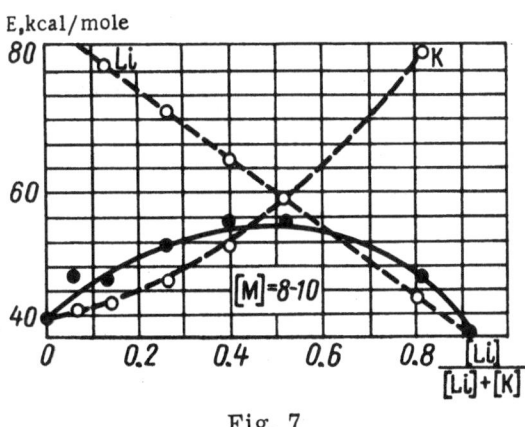

Fig. 7

We have earlier established that divalent ions in normal alkali glasses of appreciable electrical conductivity produce a decrease in the mobility of the alkali ions [17]. As a result of our work it can now be considered as established that in order to obtain the most easily fusible glass which has at the same time the lowest electrical conductivity it is necessary to include two alkali oxides in equimolar amounts in the glass. In fact in this case, as Kloster in particular has shown for borate glasses [18] and in agreement with our results, a minimum in the softening temperature coincides with a minimum in the electrical conductivity. Bearing in mind the similarity already discussed in the behavior of borate and silicate glasses with respect to their conductivity, it would be expected that the analogous property of the parallel behavior of the melting point and conductivity would also be observed in silicate glasses. It suggests that in four-component systems there will be an even clearer minimum in the conductivity with an equimolecular content of three different alkali oxides in a glass. We propose subsequently to study this technically important problem.

9. We noted earlier the unique position of the vitreous state among the other states of aggregation. It was indicated at the same time that the vitreous state, being intermediate between the liquid and the crystalline, is of particular interest from the point of view of order in the distribution of particles. In this connection the value of knowing the displacement energy of the ions in glass and the opportunity for determining it from measurements of the electrical conductivity is considerable.

Experimental results and their theoretical analysis indicate the correctness of the method chosen for studying the structure of such vitreous systems. The structure of the latter, as we have seen, has been shown by this research to be dependent on the concentration of the salt-like components, while the degree of ordering in the ion distribution increases with the concentration of the latter, approaching that of the crystalline state. The data on electrical conductivity of three-component systems provides evidence that these glasses have a specifically microheterogeneous structure.

This paper has been concerned with the problem of electrical conductivity in glasses. The conclusions require further analysis, and substantiation based on other physical and chemical properties of glass. This problem cannot be solved within the framework of this paper and will be considered further. Nonetheless it is interesting, in conclusion, to consider the principal question: how can we reconcile the optical uniformity of glass with its microheterogeneous structure?

The optical uniformity of alkali glasses can probably be explained by the high dispersivity of the alkali salt-like inclusions in the anhydride matrix. The following supports this view. With an increase in the cation electrovalence we should expect an increase in the interaction of the salt-like molecules and therefore an increase in the dimensions of the associated complexes or microcrystals.

It is consistent with this view that borate systems containing alkali earth metals do in fact produce cloudy and very opalescent glasses. This suggests that there is a decrease in the dispersivity of the alkali earth borates compared with the alkali borates. The enlargement of the associated complexes must here occur in parallel with the decrease in the radius of the alkali earth cations and must be accompanied by an increase in the tendency towards the separation of the melt into two phases and to its crystallization.

There is experimental confirmation in that the separation of alkali earth boric oxide systems is observed at contents of BaB_4O_7 of 6.85 mole %, of SrB_4O_7 of 3.34 mole %, and of CaB_4O_7 of 2.38 mole %. Dilute solutions of magnesium borate in boric anhydride could not be obtained.

Thus, the optical properties of glass are not inconsistent with its microheterogeneous structure.

By relating our results and conclusions to other properties of glass we confirm the considerable value of the electrochemical method of studying glass.

The determination of the temperature coefficient of the conductivity is most important in this respect. It enables us to calculate the displacement energy of an ion and thus to investigate, not only the strength, but also the structure of glasses.

References

1. S. Wagner, Z. Electrochem., 45:1 (1939).
2. N. W. Taylor and S. S. Sole, J. Amer. Chem. Soc., 56:1648 (1934); J. Amer. Ceram. Soc., 18:55 (1935).
3. A. Winkelmann and O. Schott, Ann. Phys. (Wied.), 51:736 (1894).
4. S. A. Shchukarev and R. L. Myuller, Zh. Fiz. Khim, 1:625 (1930); R. L. Myuller (Müller), Nature, 129:507 (1932); R. L. Myuller and B. I. Markin, Zh. Fiz. Khim; 5:1262, 1272 (1934).
5. R. L. Myuller, Zh. Fiz. Khim., 6:616 (1935).
6. W. Beetz, Ann. Phys. Chem. (Pogg.), 92:462 (1854); H. Buff, Ann. Chem. Pharm. (Lieb.), 90:257 (1854).
7. E. Warburg, Ann. Phys. (Wied.), 21:622 (1884).
8. R. Ambronn, Phys. Z., 14:112 (1913); Ann. Phys., 58:139 (1919).
9. G. Gehlhoff and M. Thomas, Z. techn. Phys., 6:544 (1925).
10. M. LeBlanc and F. Kirschbaum, Z. Physik. Chem., 72:168 (1910).
11. J. Curie, Ann. Chim. Phys., 17:6, 385 (1889).
12. V. V. Vargin, K. S. Evstropev, K. A. Krakau, I. M. Prok, and A. I. Stozharov, The Physico-Chemical Properties of Glass and their Dependence on Composition, Moscow—Leningrad (1937).
13. S. B. Thomas, J. Phys. Chem., 35:2103 (1931).
14. M. E. Spaght and J. D. Clark, J. Phys. Chem., 38:833 (1934).
15. R. L. Myuller (Müller), Phys. Z. Sowjetunion, 1:407 (1932).
16. E. Seddon, E. J. Tippett, and W. E. S. Turner, J. Soc. Glass Technol., 16:459 (1932).
17. B. I. Markin and R. L. Myuller, Zh. Fiz. Khim., 7:592 (1936).
18. H. S. Van Klooster, Z. anorg. Chem., 69:122, 135 (1911).
19. N. I. Brodskaya and V. S. Tatarinova, Uch. Zap. Leningr. Gos. Univ., No. 54, p. 249 (1940).

I

The Electrical Conductivity of Ionic-Covalent Materials*

Introduction

The physical basis for the study of the passage of electric current through dielectrics has been developed in the course of the last ten years. Many of the fundamental problems were posed, and in fact solved, here in the Soviet Union [1-3].

The physical chemical aspect of the phenomenon has received rather less attention. To a certain extent this could not but be reflected in the understanding of some of the complex phenomena produced by the chemical nature of dielectrics. At the same time as the use of physical methods enables us to find a strictly quantitative solution of the chemical problems, so a clear chemical concept concerning the material of the medium will help to establish a more valid understanding of the physical phenomena which occur in this medium.

Experimental physical chemical analysis of the electrical conductivity of ionic-covalent amorphous solids is reported in these papers. The electrochemical approach is the result of a development of the essential electrophysical concepts. In the papers which follow we shall show that the views expressed here are in good agreement with experiment and, to some extent, enlarge the circle of experimental facts amenable to theoretical generalization.

The uniqueness of the electrochemical principles thus established for ionic-covalent media is determined above all by the structural chemistry and the kinetics of processes in such media. These features are examined here on the basis of contemporary concepts concerning the structure of matter which, in the final analysis, are based on Mendeleev's Periodic System and on Butlerov's theory of the chemical structure of covalent compounds. In the first place we shall undertake a brief examination of the chemical features of various dielectric media. This is necessary so that physicists may have a correct understanding of the more recent analysis of the effect of the structural chemistry of the medium on electrolytic dissociation and ion mobility.

It is relevant to note that the structural chemical formulations used in this series of papers as characteristic of the material media in question are the result of both chemical and x-ray analyses. The structural chemical formulations are thus not arbitrarily created as models by the author, but refer in fact to inherent structural elements in the solid dielectrics in question.

The Electrochemistry of Solutions and Molten Salts

1. The fundamental principles of classical electrochemistry were established in the 19th century as a result of research on aqueous and nonaqueous solutions of electrolytes. These solutions are characterized by having, as structural units, polar molecules of a low melting, volatile solvent. A sharp decrease in the electrostatic interaction of the ions is particularly characteristic of such solutions. In spite of the high value of the dissociation energy of strong

*R. L. Myuller, Zh. Tekhn. Fiz., 25(2):236 (1955).

electrolytes (hundreds of kilocalories), they dissociate easily in a polar solvent at room temperature as a result of ion solvation, the energy of which is commensurate with the dissociation energy of the electrolytes.

The individual differences in the properties of the ions are smoothed out as a result of solvation. Their radii differ comparatively little in the solvated state, hence the closeness of the transport numbers of the ions. As a result of solvation the ion mobility in very dilute solutions is determined, in fact, by the viscosity of the solvent. The product of the specific conductivity and the coefficient of viscosity is constant and their respective temperature coefficients are found to be numerically identical but opposite in sign. The monotonic fall in molar conductivity with a rise in concentration, as a result of the association of the ions in accordance with the active mass law, is also typical of these solutions. Undissociated polar molecules of the electrolyte itself are also present as structural units in weak electrolyte solutions together with the ions and the polar molecules of the solvents.

The classical theory of electrolytic dissociation and the classical electrochemistry based on it were developed on the basis of such solutions in which polar molecules of the electrolyte and solvent, together with the ions, were the structural chemical units. The later developments of this theory, allowing for the electrostatic ion-interactions in very dilute solutions and extended to concentrated solutions by dealing with their activity, apply to these same ion–molecular solutions.

Significant deviations from the classical theory are observed with the substitution of the polar molecules of the solvent by nonpolar molecules. The solutions of electrolytes in nonpolar molecular solvents show an increase in ion association and the development (when the concentration of the latter is sufficient) of more intricate complexes consisting of ions and polar molecules of the electrolyte. A similar type of "autosolvation" of an electrolyte produces a minimum in the curve of molar conductivity against concentration. At the same time a sharp increase in molar conductivity is observed at increased concentrations. The absence of ion solvation by molecules of nonpolar solvents is based on such deviations from classical principles.

An analogous inhibition of ion solvation can also be produced in a polar solvent if the temperature of the solution approaches the critical temperature of the solvent under pressure. A similar thermal desolvation causes a sharp strengthening of ion interaction and is accompanied by the aggregation of the latter followed by the development of a solid crystalline phase.

2. A complete departure from the classical theory of electrolytic dissociation occurs on the transition from ionic–molecular solutions to purely ionic melts. In the absence of a solvent the electrostatic interaction of the thermally mobile ions in the melt occurs without hindrance and is accompanied by the emergence with time of complicated ionized complexes of altered composition. The latter, participating in the thermal agitation, produce the increased electrical conductivity of ionic melts. On a statistical time average one can find a particular type of 'autosolvated' ions in these systems. They are accompanied by a weakening of the interaction forces and thus a lowering of the potential barrier to their displacement. This leveling effect of autosolvation must clearly produce the close agreement between the values of the conductivity observed in different melts.

The Electrochemistry of Ionic Crystals

and Low Melting Amorphous Solids

1. There has been an intensive study of the electrical conductivity of ionic crystals in the last decade. The direct electrostatic interaction of the regularly packed ions excludes the participation of the vast majority of them in translational thermal motion and also determines the low conductivity and the high temperature coefficient of the conductivity. The absence of the

inhibiting effects of solvation is responsible for the predominating mobility of ions of one sign only. The considerable differentiation between the values of the conductivity of crystalline materials is another associated effect.

Ioffe was the first to note that electrical conductivity in ionic crystals requires a relatively small proportion of the ions to be displaced from their lattice points as a result of thermal vibrations [1]. The equilibrium between such "electrolytically dissociated" ions displaced into the interstices of the lattice and the ions which remain in the normal position at the lattice points obeys the normal laws of statistical thermodynamics. Dissociated ions in interstitial positions are found in translational motion when they have an activation energy E_a which is required for overcoming the potential barrier to moving from one position to the next in the interstices. From such concepts Frenkel [2] created a quantitative theory of the experimentally observed exponential dependence of specific conductivity $\varkappa$ on temperature

$$\varkappa = C \exp\left(-\frac{A}{T}\right),$$ (1)

where C and A are constants. Frenkel's theoretical expression for specific conductivity may be expressed in the form

$$\varkappa = \frac{\nu\delta^2 e^2 n_0 z^2}{6kT9\cdot10^{11}} \exp\left(-\frac{E_d + 2E_a}{2RT}\right) = 3\cdot10^{-21} n_0 z^2 \exp\left(-\frac{E_d + 2E_a}{2RT}\right)\Omega^{-1}\cdot\text{cm}^{-1},$$ (2)

where $\nu = 10^{13}$ sec^{-1} is the frequency of the thermal ionic vibrations (within an order of magnitude); $\delta \cong 3\cdot10^{-8}$ cm is the distance of a single displacement of an ion in an interstitial position; $e = 4.8\cdot10^{-10}$ absol. esu is the electron charge; $k = 1.38\cdot10^{-16}$ erg/deg is Boltzmann's constant; $T = 1000°K$; n_0 is the number of ions of a given type per cm^3 of a crystal; z is the electrovalence of these ions; and E_d is the dissociation energy of an ion.

An exchange of ions at lattice points with a resulting displacement of the vacant sites ("holes") which arise as a result of dissociation ("hole conductivity") is possible, as well as the motion of dissociated ions in the interstices ("excess ionic conductivity"). In crystals, evidently, "hole" conductivity is the more probable type [4]. Vacated sites at lattice points, or holes, can also develop as the result of an equivalent number of ions of both signs leaving the bulk of the crystal for the surface [5].

The theoretical analysis confirms that the activation energy E_a is considerably lower than the dissociation energy E_d. But contemporary theory does not provide a means of calculating the exact value of this parameter. It will be shown subsequently that the calculation of the constant C from the formula

$$C = 3\cdot10^{-21} n_0 z^2 \,\Omega^{-1}\cdot\text{cm}^{-1}$$ (3)

derived from Eq. 2 and a comparison of this value with the experimental value is very important.

The considerations above bear witness to the fact that, apart from the sharp change in the electrochemical principles with the transition from an ion-molecular to a purely ionic substance and from ionic melts to ionic crystals, there is a significant dependence on the physical state of aggregation. A similar dependence exists for liquid ionic-molecular solutions which solidify with the formation of low-melting point amorphous solids.

2. Low-melting point ionic-molecular solutions in which the solvent consists of polar molecules can be comparatively readily cooled in such a way that there is a transition into a

vitreous amorphous state. Torsional vibrational degrees of freedom are frozen in the polar
molecules on cooling through the critical temperature region. This is accompanied by the fix-
ing of the particle distribution and a cessation of the translational thermal motion of the solvated ions. At the same time the degree of electrolytic dissociation of dilute solutions of strong
electrolytes does not change noticeably. But the sharp reduction in the mobility of the solvated
ions to zero produces an abrupt decrease in the electrical conductivity. According to Kobeko et
al. [6] a 10^{15}-fold drop in conductivity is observed in this case. The insignificant conductivity
remaining at low temperatures clearly shows that quite a small number of desolvated ions
have the necessary energy to move out of the solvating shell of the polar molecules of the solvent. In this connection Kobeko and Kubshinskii established the lack of proportionality between
specific electrical conductivity and viscosity. At lower temperatures the viscosity of the frozen
solvent increases noticeably more rapidly than the specific resistance measured by the conductivity of the desolvated ions.

The Structural Kinetics and Special Electrochemical
Features of High-Melting Amorphous
Solids (Borosilicates)

1. The structural peculiarities and the atom—ion kinetics of borosilicates will be examined in terms of the dependence of the vibrational frequencies on the masses and on the radii of
the spheres of the chemical forces of atomic interactions. A compound of an element of the
second period of the Periodic Table, oxygen, with other elements in which the product of their
valencies is not less than six, is typical of the substances in question. According to Butlerov's
theory such oxides have a completely defined distribution of the chemical valence bonds between
atoms in the solid and liquid states. The structure of such substances is characterized by the
existence of a three dimensional lattice of conservative valence bonds, with the silicon or boron
atoms in the lattice interstices and the oxygen atoms forming bridges. Substances with such a
chemical structure have the greatest tendency to form glasses. Moreover, their atomic kinetic
properties do not depend on the degree of order in the structure or on the purely geometric features of crystalline lattices. The valence-bond energy and the temperature dependence of the
viscosity indicate the absence of any significant number of completely broken valence bonds. It
is possible to have "switching" (change-over) of valence bonds between atoms. This switching
is possible only when there is thermal excitation of valence vibrations. In borosilicates and
similar substances these are only observed at a fairly high temperature which has been termed
"critical" by Lebedev. At lower temperatures, normally around room temperature, the valence
bonds between the oxygen and the boron or silicon atoms become rigid as a result of the high
value of the quasi-elastic coefficients, and a number of degrees of freedom relating to the thermal valence vibrations are frozen. Thus, borosilicates and similar substances are conservative
chemical systems of valence bonded atoms with the thermal valence vibrations having relatively
enhanced excitation energies.

At both normal and slightly raised temperatures such substances may be said to be in a
stable state. The frozen state, described above, of the valence thermal oscillations, the absence
of valence bond switching, and the impossibility of any further structural changes, are typical of
this state. Such a state is maintained up to the critical temperature region which, in boron oxide for example, lies above 225-275°C and in silicon oxide above 150-300°C. Substances in the
critical temperature region (and below the melting point) may be said to be in a labile state.

In the labile state the rigidity of the chemical bonds disappears, the atomic vibrations
gradually proceed toward general disorder, valence bond switching appears, and the possibility
of some types of structural change including crystallization arises. In conditions in which

there is a virtually complete retention of all the valence chemical bonds the structure of the system becomes labile, unstable in its three-dimensional geometry and mechanically deformable.*

2. The three-dimensional lattice of conservative valence bonds is basically retained when small amounts of uni- or di-valent metal oxides are introduced into the silica and boric anhydride melts. When an alkali oxide is added to silica the number of coordinating atoms around the silicon is maintained at four. The total number of valence bonds also remains unaltered but an equivalent number of connecting valence bonds are broken between the silica lattice points:

$$
\begin{array}{c}
\quad\ \ \text{O}\quad\ \ \text{O} \qquad\qquad\qquad \text{O} \qquad\quad \text{O} \\
\quad\ \ |\quad\ \ | \qquad\qquad\qquad\ | \qquad\quad\ | \\
-\text{O}-\underset{|}{\text{Si}}-\text{O}-\underset{|}{\text{Si}}-\text{O}-\ +\text{Na}_2\text{O}\ \longrightarrow\ -\text{O}-\underset{|}{\text{Si}}-\text{O}^-\ \overset{\text{Na}^+}{\underset{\text{Na}^+}{}}\ \text{O}^--\underset{|}{\text{Si}}-\text{O}- \\
\quad\ \ |\quad\ \ | \qquad\qquad\qquad\ | \qquad\quad\ | \\
\quad\ \ \text{O}\quad\ \ \text{O} \qquad\qquad\qquad \text{O} \qquad\quad \text{O} \\
\quad\ \ |\quad\ \ | \qquad\qquad\qquad\ | \qquad\quad\ |
\end{array}
$$

The addition of alkali oxide to silica is thus accompanied by the weakening of the conservative character of the system and the structure, as a whole, disintegrates. At the same time there is an increase in the number of labile electrostatic bonds with low quasi-elastic coefficients between the metal ions and the ionized oxygen atoms which retain a single bond with the covalent lattice. In silicates this is accompanied by a decrease in the lower limit of the critical temperature region and of the melting point, and crystallization is facilitated.

The addition of alkali oxide to boron anhydride has a slightly different effect. In this case the number of coordinating oxygen atoms around the boron increases from three to four, the six-electron valence shell of the central boron atom changes to the more stable eight-electron, and the number of connecting valence bonds in the structure increases:

$$
\begin{array}{c}
\qquad\qquad\qquad\qquad\qquad\qquad\qquad\qquad\ \text{O}\quad\ \ \text{O} \\
\qquad\qquad\qquad\qquad\qquad\qquad\qquad\qquad\ |\quad\ \ | \\
-\text{O}-\text{B}-\!\!\!\begin{array}{c}\diagup\text{O}\diagdown\\ \diagdown\text{O}\diagup\end{array}\!\!\!+\text{Na}_2\text{O}+\!\!\!\begin{array}{c}\diagdown\text{O}\diagup\\ \diagup\text{O}\diagdown\end{array}\!\!\!\text{B}-\text{O}-\ \longrightarrow\ -\text{O}-\underset{|}{\text{B}}^-\!\overset{\text{Na}^+}{-}\text{O}-\underset{|}{\text{B}}^-\!\overset{\text{Na}^+}{-}\text{O}- \\
\qquad\qquad\qquad\qquad\qquad\qquad\qquad\qquad\ |\quad\ \ | \\
\qquad\qquad\qquad\qquad\qquad\qquad\qquad\qquad\ \text{O}\quad\ \ \text{O} \\
\qquad\qquad\qquad\qquad\qquad\qquad\qquad\qquad\ |\quad\ \ |
\end{array}
$$

In this case the boron atoms then become negatively ionized. On the whole, in borates there is an increase in the stability of the chemical structure which is strengthened and the lower limit of the critical temperature region is raised. The simultaneous accumulation of labile ionic groupings is relevant to the crystallization processes.

Thus, alkali oxides loosen the structure of silicates but strengthen that of borates. The Zachariasen—Warren theory is thus mistaken in the assumption that the initial addition of these oxides to boric anhydride breaks valance bonds.

Oxides of alkali-earth metals increase the strength of the structure in borosilicates due to the positioning of the divalent cations of the oxygen atoms in sites left by opened valence bonds.†

*R. L. Myuller determines the critical temperature region in the case of the simplest glasses from data on their thermal capacity (see notes on published literature, section V, at end of book). In the case of more complicated glasses he uses the results of a physical chemical analysis. In that instance the region of critical temperature, according to Myuller, does in fact coincide with the so-called "anomalous range of temperatures" (see, for example; K. S. Evstrop'ev, and N. A. Toropov, The Chemistry of Silicon and the Physical Chemistry of Silicates, Promstroiizdat (1956). (Editor's note)

†When oxides of alkali-earth metals are introduced into silicate and borosilicate glasses a complex change in the structure of the glasses occurs and the strength of the structure (defined according to Myuller by the value of the critical temperature) may either increase or decrease depending on the size of the alkali-earth metal ion and the original glass composition. (Editor's note)

Silicates and borates thus contain the purely covalent nonpolar components of a continuous structural network $(SiO_{4/2})_n$ and $(BO_{2/2})_n$ and also negatively ionized parts $(SiO_{3/2}O^-)_n$ and $(B^-O_{4/2})_n$. They are conservatively interconnected in a continuous covalent network. The valence bonds in them are strong, not only in the solid state but also the liquid state. Silicates and borates (not too saturated with metallic oxides) have, therefore, scarcely any mobile anions and naturally motion in them is due to cations. Ostroumov proved experimentally that, within 2%, the cation transport number in vitreous borax was unity [7]. The assumption that both oxygen and silicon ions migrate, not only in the solid but also in the melt, is clearly mistaken.

Cations in borates are coupled to the negatively ionized boron atoms by fairly electrostatic bonds, with an energy commensurate with the energy of single valence bonds (of the order of 100 kcal/mole). As we have shown already, the labile nature of ionic bonds is the result of the value of the quasi-elastic coefficient being lower than in the case of covalently bonded atoms. This in itself can give a relative indication about the cation bonds with the ionized oxygen atoms in silicates.

3. From these considerations it is clear that the continuous covalent structural network in borates and silicates consists essentially of two types of structural elements: nonpolar $BO_{3/2}$ and $SiO_{4/2}$, and polar $B^-O_{4/2}$ and $SiO_{3/2}O^-$. Their quantitative relationship and volume concentration in the glass determines the physical properties of vitreous high-melting substances, and in particular, the electrical conductivity, dielectric constant, dielectric loss, viscosity, etc.

The structural units have free valencies which link them in a continuous, strong, spatially unlimited chemical structure. Clearly in such systems there can be no molecular structures with weak van der Waals interaction forces. It is therefore meaningless to use molecular type formulae to express the composition of such substances.

We must point out here one special feature of these polar structural elements. Oxygen atoms in tetrahedral coordination around the boron or silicon atoms form four potential wells in the spaces between themselves. In borates, a univalent cation coupled to the central ionized boron atom is situated in one of these. In silicates, univalent cations coupled to the ionized oxygen atoms lie nearby in one of three potential wells. Thus, under the influence of an external field or in the presence of a dissociated mobile cation, there must occur in the matrix of the polar structural elements a displacement of univalent cations proportional to the external or local ionic field. Such a displacement is activated but the activation energy will be insignificant.

On the whole, the phenomenon of univalent cation displacement within the polar structural elements can be related to the orientation of polar molecules in ionic-molecular solutions. A dissociated cation in a matrix of polar structural units which contain univalent cations must be subject to the effect of the surrounding medium which is an identical effect to the autosolvation of an electrolyte by polar molecules in a nonpolar medium. Divalent cations, directly bonded electrically to two negatively ionized structural elements, cannot produce such an orientation effect.

This phenomenon of effective autosolvation makes the concentration dependence of electrical conductivity in glasses analogous to the dependence observed in solutions of electrolytes in nonpolar molecular solvents. The curve of molar conductivity against electrolyte concentration (equivalent to the total concentration in the glass of univalent cations, dissociated and non-dissociated) would be expected to show a minimum followed by a sharp rise. This feature is also seen in the molar conductivity of silicate* and borate glasses containing oxides of univalent

---

*Because of the considerable experimental difficulty this principle has not yet been demonstrated in the case of two-component alkali silicate glasses. Myuller based his conclusions chiefly on his careful studies of the concentration dependence in alkali borate glasses. (Editor's note)

metals [8]. The author has already indicated the probability of effective autosolvation under these conditions [9]. This purely qualitative exposition of the phenomenon requires, of course, an experimental foundation and this will be the aim of the following papers.

4. The conservative covalent structure in the liquid and, more particularly, in the labile vitreous state, is modified very slowly as the temperature is reduced. As a result, a thermodynamically nonequilibrium vitreous state is obtained. In the stable state the frozen skeleton of valence bonds with relatively weaker thermal vibrations of a degraded type, cannot quite be modified to accord with the temperature.

The equilibrium electrolytic cation dissociation is a different matter. The thermally labile cations, participating fully in the distribution of energy among the degrees of freedom, are distributed in a strict thermodynamic equilibrium between ionized and nonionized structural elements. In the stable state of glass electrolytic dissociation is not accompanied by any perceptible modification of the skeletal structure which has rigid valence bonds. In the labile state, however, the covalent skeleton may, during electrolytic dissociation, undergo some local structural breakdown producing an entropy change.

The electrical conductivity of glasses with a low alkali content is, of course, determined by the movement of dissociated cations in a nonpolar anhydride medium [10]. In alkali-rich glasses the conductivity mechanism will involve the displacement of vacant, electrically negative "holes" in the compact medium of the associated polar structural elements. These "holes" in glass can, according to Frenkel, arise from the electrolytic dissociation of cations. The appearance of defects, the result according to Schottky of the emergence at the surface of the glass of an equal number of changed particles of both signs, i.e. of whole polar structural units, is unlikely. This is because of the high energy for the simultaneous disruption of quadrivalent bonds (of an order of 400 kcal/mole) which is considerably greater than the electrolytic dissociation energy when univalent cations move into the intralattice space (of the order of 50 kcal/mole). The fact that the cation transport number is one, supports the view that the defects in glass are of the Frenkel type and that only a very small number of Shottky defects are present. A further indication is the considerable magnitude of the activation energy of viscosity which is approximately twice that of the electrical conductivity in the melt [12].

There is no direct physical connection between the viscosity of molten glasses and their conductivity.* According to the author's theoretical analysis, the viscosity of aluminoborosilicate glasses is produced in the first instance by the content of the covalent component and is directly proportional to the packing density of the strong valence bonds. Conductivity is proportional to the cation content of the glass and is determined by the strength of the labile electrostatic bond with the ionized covalent skeleton of the glass. In fact, according to Evstrop'ev, the product of the coefficient of viscosity and the specific conductivity is not a constant and the temperature coefficient of the first parameter exceeds the temperature coefficient of the second [13].

5. A brief review of the structural and kinetic features of ionic-covalent borosilicates and some notes of a qualitative character concerning the electrochemistry of such substances makes it possible to examine from a new standpoint the quantitative principles of conductivity and closely related phenomena.

Conclusions

1. The electrochemistry of a) ionic-molecular; b) purely ionic; and c) ionic-covalent systems, must be differentiated according to the structural chemistry and state of aggregation of the medium.

_____
* This is dealt with in more detail in another paper by R. L. Myuller starting on p. 121 (Editor's note).

2. In borosilicates the electronegatively ionized atoms are linked by valence bonds with the conservative covalent lattice and do not produce mobile anions.

3. The structural elements in borosilicates consist of polar and nonpolar trigonal or tetrahedral groups, with the atoms bound together within the groups by valence bonds. These groups, $BO_{3/2}$, $SiO_{4/2}$, $B^-O_{4/2}M^+$, and $SiO_{3/2}O^-M^+$, have three or four externally directed free valence bonds by which they are interconnected into the strong, continuous network.

4. The thermodynamic equilibrium state of the electrolytic dissociation in borosilicates occurs as the result of a small proportion of the cations leaving the polar structural elements for the interstitial space. In this case there is an autosolvation-type effect involving the interaction of dissociated cations with the polar structural elements surrounding them.

5. The autosolvation effect is due to the oriented displacement of univalent cations in polar structural elements. Such orientation must be absent with divalent cations.

6. Electrolytic dissociation in borosilicates in the critical temperature region must be accompanied by a change in entropy as a result of the local changes in what, under these conditions, is a labile covalent structural network.

7. The mechanism by which motion of the dissociated cations occurs in borosilicates is analogous to that observed in solid ionic crystals. With a sufficiently high content of polar structural units the conductivity is most probably of the "hole" type.

8. The development of structural defects of the Schottky type in the ionized covalent network is unlikely in glasses.

9. There is no direct connection between viscosity and electrical conductivity in borosilicates since the first parameter has a linear dependence on the valence-bond packing density, while the second is determined by the concentration of cations and by the strength of their bond with the ionized covalent network.

10. Ionic-covalent substances, which are clearly distinct from ionic solutions in polar solvents (liquid and solid), are in their electrochemical properties (slight electrolytic dissociation with comparatively high mobility of the dissociated cations) close to solid ionic crystals. In their concentration dependence of conductivity these substances have a community with liquid solutions of electrolytes in nonpolar solvents.

11. The specific feature that distinguish ionic-covalent systems from all other systems is a ubiquitous purely cationic conductivity.

12. The qualitative features of electrochemical phenomena in borosilicates discussed here require a further quantitative experimental and theoretical foundation.

References

1.   A. F. Ioffe, The Elastic and Electrical Properties of Quartz, SPb (1915); Crystal Physics, Gosizdat, Moscow—Leningrad (1929); A. F. Val'ter (Ed.), Physics of Dielectrics, Gosizdat, Moscow—Leningrad (1933).
2.   Ya. I. Frenkel', Z. Phys., 35:652 (1926); Ya. I. Frenkel', A Kinetic Theory of Liquids, Izd. Akad. Nauk SSSR, (1945).
3.   G. I. Skanavi, Physics of Dielectrics, GITTL, Moscow—Leningrad (1949).
4.   W. Schottky, Z. Physik. Chem., 29:335 (1935).
5.   C. Wagner and W. Schottky, Z. Physik. Chem., 11:163 (1939).
6.   P. P. Kobeko, The Amorphous State, Akad. Nauk SSSR, Moscow—Leningrad (1936); P. P. Kobeko and E. Kuvshinskii, Zh. Tekhn. Fiz., 8:608 (1938).
7.   G. Ostroumov, Zh. Obshch. Khim., 19:407 (1949).

8. S. A. Shchukarev and R. L. Myuller, Zh. Fiz. Khim., 1:625 (1930); S. A. Shchukarev (Schtschukarew) and R. L. Myuller (Müller), Z. Physik. Chem., A150(5/6):439 (1930); R.L. Myuller and B. I. Markin, Zh. Fiz. Khim., 5(9):1272 (1934); R. L. Myuller (Müller) and B. I. Markin, Acta Physicochim. URSS, 1:266 (1934); ●R. L. Myuller, Uch. Zap. Leningr. Gos. Univ., No. 54, p.159 (1940); Izv. Akad. Nauk SSSR, Ser. Fiz., 4:607 (1940) ● this volume, p. 3 and p. 170 resp.

9. R. L. Myuller, Zh. Fiz. Khim., 6:616 (1935); R. L. Myuller (Müller), Acta Physicochim. URSS, 2:103 (1935).

10. R. L. Myuller (Müller), Phys. Z. Sowjetunion, 1:407, 1932; R. L. Myuller, Nature, 129:507 (1932).

11. N. Mott and R. Gurney, Electron Processes in Ionic Crystals, Oxford Clarendon Press (1940).

12. A. M. Samarin and L. A. Shvartsman, Usp. Khim., 21:336 (1952).

13. I. V. Grebenshchikova (Ed.), Physico-chemical Properties of the Three-Component System, $Na_2O-PbO-SiO_2$, Izd. Akad. Nauk SSSR, Moscow–Leningrad (1949); K. S. Evstrop'ev and N. A. Toropov, The Chemistry of Silicon and The Physical Chemistry of Silicates, Promstroizdat, Moscow (1950).

II

Experimental and Theoretical Expression for
the Molar Electrical Conductivity of Borosilicates*

Introduction

1. The slight electrolytic dissociation, the comparatively high mobility of the dissociated ions, the unipolar conductivity, and the solid state — all these connect vitreous borosilicates, in an electrochemical context, with ionic crystals. Starting from this point, the author made a comparison between experimental results on the electrical conductivity of glasses and the values calculated from Frenkel's theoretical formula [1]. It was clearly necessary to have a physical chemical justification for the application of this to amorphous, ionic-covalent glasses, since the formula had been derived for purely ionic crystalline solids. A regular lattice is absent in these glasses and they have, in addition to the conduction cations, immobile negatively ionized and nonionized atoms mutually interconnected by valence bonds in a three-dimensional network structure.

Because of the need for a calculation of the number of vacant sites for dissociated cations in a nonpolar glass medium the author has produced a statistical equation for the conductivity of glasses. At the conference on the vitreous state held in Leningrad in 1939, we presented a preparatory communication concerning a rigid basis for an equation applicable to ion-covalent systems [2]. In this paper we present the resultant rigidly based theoretical equation for the conductivity of borosilicates, with appropriate changes to make the resulting formula more precise. At the same time a theoretically based transformation of the molar conductivity equation to include glasses in the labile state (in the critical temperature region) is suggested.

2. It was emphasized in the previous communication that cations in borosilicates are electrostatically associated with the negatively ionized boron and oxygen atoms of the covalent network. These electrostatic bonds have considerable energy, around 50 kcal/mole [3]. This produces a very low degree of electrolytic dissociation in borosilicates. The insignificance of the dissociation becomes clear if the values of the electrical conductivity are directly compared in two glasses; one with a very low and one with a very high alkali oxide content. Table 1 shows an example of the experimental results for two sodium borate glasses [4]. The value of the conductivity $\varkappa$ and the temperature are given in Table 1 as well as the molar sodium concentration: n is the number of sodium ions per ml; $\vec{n}$ is the number of sodium ions passing through a cross section of 1 cm^2 in 1 sec with a field of 1 V/cm; δn is the number of sodium ions in a monoionic layer ($\delta \cong 2 \cdot 10^{-8}$ cm) per cm^2 of the cross section of the glass.

It is clear that $\vec{n} = \varkappa / e$ of ions/cm^2. sec, where e the cation change is equal to $1.6 \cdot 10^{-19}$C.

It follows from these results that in glass with a low alkali oxide content a vanishingly small percentage of all the ions present, only $1.8 \cdot 10^{-7}\%$ in all, are set in translational motion

*R. L. Myuller, Zh. Tekhn. Fiz., 25(2):246 (1955).

TABLE 1

Glass No.	Sodium ion concentration mole/ml	n	δn	T, °C	$\varkappa$, $\Omega^{-1} \cdot cm^{-1}$	$\overrightarrow{n}$ ions/cm²·sec	% of ions displaced in 1 sec
I	$5.8 \cdot 10^{-6}$	$3.2 \cdot 10^{18}$	$7.0 \cdot 10^{10}$	220	$2.0 \cdot 10^{-17}$	125	$1.8 \cdot 10^{-7}$
II	$2.3 \cdot 10^{-2}$	$1.4 \cdot 10^{22}$	$2.8 \cdot 10^{14}$	293	$1.5 \cdot 10^{-5}$	$9.4 \cdot 10^{13}$	34

per second. In alkali-enriched glass 34% per second of all the ions can be successfully displaced. In other words, with a change in temperature and cation concentration, an increase of $2 \cdot 10^8$ times in the percentage of displaced ions is observed in sodium borate glasses.

It is quite obvious that we cannot attribute such a considerable increase in ion displacement to only one mechanical factor, namely the mobility of the ions in the medium. It is accepted also that we must here differentiate between highly mobile and immobile ions. In other words, the results clearly indicate that there is electrolytic dissociation of the polar structural units $B^-O_{4/2}M^+$, with a fairly high dissociation energy which decreases significantly with the concentration of alkali oxide in the glass.

In contrast to the electrolytic dissociation energy, the activation energy of the displacement of a dissociated cation is insignificant and must be only a few kcal per mole [3]. The deciding factor in the conductivity of glasses is, as the experimental results [4, 5] show, the appreciable energy, basically that of the electrolytic dissociation of the cation as it is detached from the polar structural group and enters the intranetwork space.

The Molar Electrical Conductivity of Glasses

in the Stable State

1. Electrolytic dissociation in thermodynamic equilibrium occurs in borosilicates when the cations leave the polar structural elements $B^-O_{4/2}M^+$ or $SiO_{3/2}O^-M^+$ and move into the intranetwork space in the matrix of the polar groups and of the nonpolar structural units $BO_{3/2}$ and $SiO_{4/2}$. With a low metal oxide content the conductivity of glass is limited by the increased potential energy for the movement of dissociated cations in the nonpolar medium (Fig. 1). With a sufficiently high metal oxide content the movement of the mass of dissociated cations occurs in the medium of the associated polar structural units. We must thus examine the two equilibria of electrolytic dissociation. Their coexistence does not complicate the analysis of the phenomenon, since in both cases the vast majority of the polar groups remains undissociated and therefore, to a first approximation, the concentration of nondissociated polar groups can be equated to their total concentration.

2. Electrolytic dissociation, with a cation transition into the medium of the nonpolar structural units, occurs in the borsilicates and silicates according to the chemical equilibrium equations,

$$\left.\begin{array}{l} B^-O_{4/2}M^+ + BO_{3/2} \rightleftarrows B^-O_{4/2} + M^+BO_{3/2}, \\ SiO_{3/2}O^-M^+ + SiO_{4/2} \rightleftarrows SiO_{3/2}O^- + M^+SiO_{4/2}. \\ \phantom{SiO_{3/2}O^-M^+}_{(n-n_d)} \phantom{+ SiO_{4/2}}_{(N-n_d)} \phantom{\rightleftarrows SiO_{3/2}O^-}_{n_d} _{n_d} \end{array}\right\} \qquad (1)$$

Dissociated cations (n_d is the cation concentration per ml) are distributed among the nonpolar structural units and are connected to these by relatively weak polarization forces. It is considered in this case that a nonpolar structural unit does not retain more than one cation, due both to the electrostatic repulsion of like charged ions and to their very low concentration.

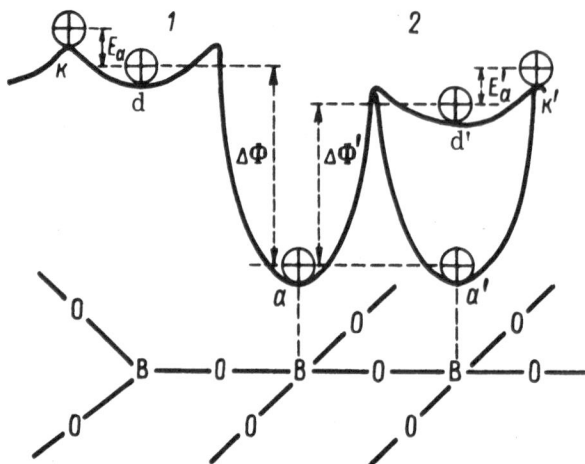

Fig. 1.   1) nonpolar medium;   2) polar medium.

If the total number of cations (equal to the number of polar structural elements) both dissociated and undissociated is n per ml, then $n - n_d$ is the number of undissociated polar groups. If the total number of nonpolar structural elements is N per ml then $N - n_d$ is the number of unoccupied nonpolar units not connected to dissociated cations. On dissociation, a cation moves from a potential well a near the polar structural element into a potential well d near the nonpolar structural unit (see Fig. 1). The difference in the mean statistical levels of potential energy will correspond to the change in the standard free energy of electrical dissociation $\Delta\Phi$. Because of the negligible concentration of dissociated cations the interaction between them and the residual vacant negatively charged structural groups $B^-O_{4/2}$ or $SiO_{3/2}O^-$ is, in fact, absent. For the same reason the nature of the interaction between the polar and the nonpolar structural units does not change. It is therefore possible to apply here the mass action law for extremely dilute solutions using concentration rather than activity. Then we obtain, from Eq. (1), the thermodynamic equality

$$\ln \frac{n_d^2}{(n - n_d)(N - n_d)} = -\frac{\Delta\Phi}{RT}.\tag{2}$$

Taking into account the marked displacement of Eq. (1) to the left-hand side, giving $n \gg n_d \ll N$, we obtain the expression for the number of ions in 1 cm³ of glass:

$$n_d = \sqrt{nN} \exp\left(-\frac{\Delta\Phi}{2RT}\right).\tag{3}$$

3. The free energy $\Delta\Phi'$ of cation displacement as a result of electrolytic dissociation into the environment of the polar structural elements, must be slightly lowered because of the lower value of the potential energy of dissociated cations (in potential wells d' Fig. 1). As has already been indicated the latter is produced by the autosolvation effect of oriented cation displacement in the surrounding polar groups [3]. Each polar structural unit $SiO_{3/2}O^-M^+$ or $B^-O_{4/2}M^+$ can absorb one dissociated cation (obtaining complexes $M^+SiO_{3/2}O^-M^+$ or $M^+B^-O_{4/2}M^+$) and only one, owing to the repulsion of like charged cations and their low concentration in the dissociated state. Thus, electrolytic dissociation in a polar medium is described by the chemical equations

$$2\,(B^-O_{4/2}M^+) \rightleftarrows B^-O_{4/2} + M^+B^-O_{4/2}M^+,$$
$$\underset{2\,(n - n_d')}{2\,(SiO_{3/2}O^-M^+)} \rightleftarrows \underset{n_d'}{SiO_{3/2}O^-} + \underset{n_d'}{M^+SiO_{3/2}O^-M^+}.\tag{4}$$

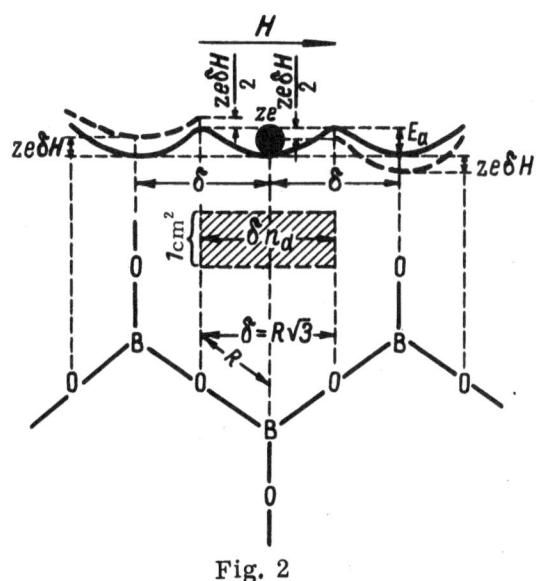

Fig. 2

4. From chemical structural data on borosilicates, dissociated cations in a nonpolar medium must be connected to the nonpolar structural units by weak polarization forces which represent the simultaneous interaction at d of the boron and two or three oxygen atoms (Fig. 1). When there is an activation energy E_a, such cations with isotropic disordered vibrational motion can overcome the small potential barrier to movement into the neighboring potential well. The activation energy for the displacement is necessary to rupture the polarization bond with the boron and the one or two oxygen atoms. This is shown diagramatically in Fig. 2 which corresponds to established physicochemical data [6].

In the absence of an external electrical field such displacements have a disordered thermal character. An external field adds directionality to this motion. The translational directed cation current is directly proportional to the field at low fields.

Figure 2 shows diagramatically the sequential potential wells along the field. A dissociated cation with a charge ze is centrally situated in them. In the absence of a field the activation energy E_a of cation displacement is identical in both directions (continuous potential curve). By applying an external field H the levels of cation potential energy in neighboring wells are displaced by $ze\delta F$ which corresponds to the drop in the electrostatic potential H in the interval δ between the wells. The displacement of the potential curves (broken line in Fig. 2) results, according to Polani, in a change in the value of the activation energy of the cation displacements by a value $\pm 0.5\, ze\delta F$.*

Hence, it is easy to obtain an expression for the resulting number of dissociated cations displaced in the direction of the external field through 1 cm² of the cross-section of 1 sec in a field of 1 V/cm:

$$\vec{n} = \frac{\delta^2 n_d \nu ze}{3kT} \exp\left(-\frac{E_a}{RT}\right),\ ^\dagger \tag{5}$$

where ν is the thermal vibrational frequency; k is the Boltzmann constant.

5. The equation for the specific conductivity $\varkappa$, limited to activated dissociated cations in a nonpolar glass matrix, is obtained by substituting Eq. (3) in Eq. (5):

---

*The coefficient is 0.5 because the distance between the potential well and the potential barrier is half the distance δ between the two wells. (Editor's note)

†In the original of this paper, and in some of the later ones, the numerical coefficient of the pre-exponential factor was inaccurate but it was corrected in a later series of papers by the author. The expression had the form

$$\vec{n} = \frac{\delta^2 n_d \nu ze}{12kT} \exp\left(-\frac{E_a}{RT}\right).$$

Since, however, the pre-exponential factor can only be calculated within an order of magnitude this inaccuracy did not significantly affect the author's conclusions. The appropriate corrections have been made in all the papers in this selection. (Editor's note)

$$ \varkappa = \frac{i}{H} = \frac{\overrightarrow{nze}}{H} = \frac{\delta^2 v z^2 e^2}{3kT} \sqrt{nN} \exp\left(-\frac{\Delta\Phi + 2E_a}{2RT}\right), \tag{6} $$

where i is the current density.

In the case of a glass matrix of polar structural elements the value of $\varkappa$ can be found by substituting Eq. (4) in Eq. (5)

$$ \varkappa = \frac{\overrightarrow{nze}}{H} = \frac{\delta^2 v z^2 e^2}{3kT} n \exp\left(-\frac{\Delta\Phi' + 2E_a'}{2RT}\right). \tag{7} $$

6. From the structural data for the boric anhydride medium we can determine the size of the individual displacements of a cation $\delta = R\sqrt{3}$, where R is the sum of the atomic radii* of boron and oxygen (see Fig. 2). According to x-ray results, R = 1.36 Å [18] in boric anhydride and thus δ = 2.4 Å. In a borate medium the oxygen atoms are in a tetrahedral coordination and the parameter corresponding to the distance between the centers of the oxygen atoms will be 1.63 R. According to the x-ray data, R = 1.53 Å [8] in this case and therefore δ = 2.5 Å. Such a calculation makes no allowance for a change in the orientation of the structural units with respect to the external field. The calculation, however, of the mean statistical value of the parameter does not have a decisive significance since the value of the frequency $\nu = 10^{13}\,\mathrm{sec}^{-1}$ in the equation is accurate only to an order of magnitude.

By substituting the calculated values of the parameters in Eq. (6) and by changing to molar concentration, we obtain (assuming T = $5 \cdot 10^2\,^\circ\mathrm{K}$)

$$ \varkappa = 4.4 \cdot 10^8 z^2 \sqrt{[M][\zeta]} \exp\left(-\frac{\Delta\Phi + 2E_a}{2RT}\right) \Omega^{-1} \cdot \mathrm{cm}^{-1}, \tag{6a} $$

where [M] and $[\zeta]$ are, respectively, the concentrations of cations and nonpolar structural units expressed in mole/ml. The last expression is transformed without difficulty into an equation for molar conductivity (mobile cations in a nonpolar medium)

$$ \Lambda = 4.4 \cdot 10^8 \sqrt{\gamma z^5} \exp\left(-\frac{\Psi_\Phi}{2RT}\right) \Omega^{-1} \cdot \mathrm{cm}^{-1} /\mathrm{mole}, \tag{8} $$

where $\gamma = \dfrac{[\zeta]}{z[M]} = \dfrac{[\zeta]}{[\zeta^-]}$ corresponds to the molar ratio in the glass of nonpolar $([\zeta]\,\mathrm{mole/ml})$ to polar structural units $([\zeta^-]\,\mathrm{mole/ml})$; $\Psi_\Phi = \Delta\Phi + 2E_a$ is an energy, close in fact to the value of the free energy of dissociation of the cations since it is known that $E \ll \Delta\Phi$ [3].

We can obtain a completely analogous equation for the molar conductivity determined by mobile cations in a polar medium in borate glasses:

$$ \Lambda = 4.4 \cdot 10^8 z^2 \exp\left(-\frac{\Psi_\Phi}{2RT}\right) \Omega^{-1} \cdot \mathrm{cm}^{-1} / \mathrm{mole}. \tag{9} $$

7. In silicate glasses the tetrahedral atomic radii of silicon and oxygen are, respectively, 1.17 and 0.66 Å [8], and the distance between the oxygen atoms $\delta = 1.63$ R = 3.0 Å (R is the sum of the atomic radii of silicon and oxygen). Thus, in changing from borate to silicate glasses, the pre-exponential term in Eqs. (8) and (9) must increase 1.56 times ($4.4 \cdot 10^8$ instead of $6.8 \cdot 10^8$). This, however, has no real significance since, as already noted, the derivation was only to an order of magnitude.

In a polar medium it is more likely that conductivity involves mobile vacant sites produced as a result of the displacement of cations from the neighboring polar structural groups [3]. In this case the free energy of dissociation remains equal to the energy for cation production, while the activation energy possibly becomes somewhat lower. The preexponential factor in Eq. (9) is retained unchanged and on the whole Eq. (9) does not change in form.

8. The impossibility of structural changes in the rigid covalent network of glasses in the stable state [3] leads to the absence of entropy changes on electrolytic dissociation. In stable glasses this defines the dependence of the free energy of electrolytic dissociation on temperature:

$$\frac{\partial (\Delta \Phi_\sigma)}{\partial T} = -\Delta S_\sigma = 0, \tag{10}$$

where the index σ refers to the stability of the covalent system.

Assuming that the activation energy also remains unchanged (activation entropy $\Delta S = 0$) we can predict the constancy of the energy parameter of any glass in the stable state temperature region:

$$\Psi_{\Phi \, \text{state}}^{\text{stable}} \neq f(T). \tag{11}$$

The Molar Conductivity of Glass in the Labile State

1. In order to compare numerous experimental results on the conductivity of borosilicates with the values resulting from the formulae above, we must reorganize the latter. In the first place it is necessary to make the expression for the molar conductivity of borosilicates in the labile state more precise.

It was noted in the preceding paper that electrolytic dissociation in borosilicates in the labile state must be accompanied by an entropy change caused by the occurrence of local structural changes. These occur during the reorganization of the polar structural unit into an electrically negative, ionized structural unit plus a dissociated cation. This restructuring of the local electrostatic fields polarizing the matrix must produce small local structural changes [3].

Thus, in glasses in the labile state, the free energy of cation dissociation must depend on temperature:

$$\frac{\partial (\Delta \Phi)}{\partial T} = -\Delta S \neq 0. \tag{12}$$

Compared with the stable state, in which the free energy $\Delta \Phi$ and the enthalpy ΔH of electrolytic dissociation are equal since $\Delta S = 0$, these parameters are here different and are given by the fundamental thermodynamic relationship:

$$\Delta \Phi = \Delta H - T \Delta S. \tag{13}$$

By substituting the latter in the expression for the specific conductivity (nonpolar medium) Eq. (6a) and changing to molar conductivity, we obtain

$$\Lambda = 4.4 \cdot 10^3 \sqrt{\gamma z^5} \exp \left(\frac{\Delta S}{2R} \right) \exp \left(-\frac{\Psi_H}{2RT} \right) \Omega^{-1} \cdot \text{cm}^{-1} / \text{mole}, \tag{14}$$

where $\Psi_H = \Delta H + 2E_a$ is the heat parameter of the conductivity of glasses in the labile state.

In the same way the equation for the molar conductivity in the polar medium of borosilicate glasses is transformed thus:

$$\Lambda = 4.4 \cdot 10^3 z^2 \exp \left(\frac{\Delta S}{2R} \right) \exp \left(-\frac{\Psi_H}{2RT} \right) \Omega^{-1} \cdot \text{cm}^{-1} / \text{mole}. \tag{15}$$

2. The essential feature of Eqs. (14) and (15) which distinguish them from Eqs (8) and (9) is the appearance of an entropy term in the factor multiplying the exponential temperature dependence function. This explains the abrupt change in the temperature coefficient of conductivity observed in the conductivity of glasses on the transition from the stable to the labile state. The appearance of local structural changes accompanying electrolytic dissociation is the source of such a change. The free energy of dissociation, which does not in the stable state depend on temperature, in this case decreases in the critical temperature region.

Frenkel' had already discussed the possibility of a decrease in the energy parameter Ψ_Φ in the critical temperature region. He presented an empirical equation for this effect,

$$\Psi = \Psi_0 - 2\alpha RT(T - T_0),$$

where α is a constant and T_0 is the lower limit of the critical temperature region [2].

Fundamental Experimental and Theoretical Relationships

1. As is well known a linear dependence of the specific conductivity on the reciprocal of temperature has been observed experimentally in glasses

$$\ln \varkappa = -\frac{A}{T} + B, \tag{16}$$

where A and B are constants.

An empirical equation for the molar conductivity of glass emerges from this equation

$$\Lambda_e = \frac{\varkappa}{M} = 10^{0.4343B - \log[M]} \exp\left(-\frac{A}{T}\right) \Omega^{-1} \cdot \text{cm}^{-1}/\text{mole}. \tag{17}$$

The theoretical expression for the molar conductivity is conveniently expressed in a form analogous to Eq. (17)

$$\Lambda_t = 10^{P_t} \exp\left(-\frac{\Psi_\Phi}{2RT}\right) \Omega^{-1} \cdot \text{cm}^{-1}/\text{mole}. \tag{18}$$

The comparison of the experimental and theoretical expressions is reduced to a verification of the equality

$$P_t = P_c = 0.4343 B - \log[M], \tag{19}$$

$$\Psi_\Phi = 2AR. \tag{20}$$

It now simply depends on the pre-exponential factor in Eq. (18). It is calculated theoretically quite easily for glasses in the stable state to an accuracy of an order of magnitude. This parameter is unchanged in the labile state but is supplemented by the entropy factor which enables us to determine the change in entropy and energy of dissociation of glasses in this state.

Table 2 summarizes the values of the parameters P_t and P_e based on Eqs. (8), (9), (14), (15) and (18) for polar and nonpolar media in glasses in both the stable and labile state. In the same formulae (Eqs. (8), (9), (14), and (15)) the modification on a change from borates to silicates ($\alpha = 3.64$ in borates and $\alpha = 3.83$ in silicates) should be noted.

The value of the energy Ψ_Φ in stable glasses is determined from Eq. (20).

2. In the case of stable glasses we are mainly interested in a direct comparison of the theoretical values P_t (Table 2) with the experimental P_e, using Eq. (19). In the case of labile glasses it is essential to determine the entropy of electrolytic dissociation

$$\left. \begin{array}{l} \Delta S = 4.6R\,(P_c - P_t), \\ \Psi_H = 2AR, \end{array} \right\} \tag{21}$$

TABLE 2

Physical state of glass	Dissociated cations in a matrix which is:	
	Nonpolar	Polar
Stable	$P_t = a + 2.5 \log z + 0.5 \log \gamma = P_\sigma^*$	$P_t = a + 2 \log z = P_{\sigma_p}^*$
Labile	$P_c = a + 2.5 \log z + 0.5 \log \gamma + \dfrac{\Delta S}{4.6R}$	$P_c = a + 2 \log z + \dfrac{\Delta S}{4.6R}$

* Henceforward P_σ signifies P_t calculated for a nonpolar maxtrix while $P_{\sigma t}$ is the same parameter for a polar matrix.

where P_t is the theoretical value in the stable state from Table 2. The value of the parameter Ψ_H follows from a comparison of Eqs. (14) and (17). With such values the energy Ψ_Φ is determined from the value of the enthalpy:

$$\Psi_\Phi = \Psi_H - T\Delta S. \qquad (22)$$

The relatively small value of the activation energy E_a allows us to estimate, to a first approximation, the value of the free energy of dissociation $\Delta\Phi$ for labile glass from the value of Ψ_Φ.

3. This method of calculating the thermodynamic dissociation parameters in labile glasses assumes a constancy in the coefficients A and B in the critical temperature region.

The latter is only valid over a narrow region of temperature and only as a rough approximation. The determination of the variable parameter $\Psi_\Phi(T)$ directly from the general equation (18) is more valid as it takes into account the effective constancy of the statistically calculated parameter P_t when the glass is in the labile state:

$$\Psi_\Phi(T) = 4.6R(P_t - \log\Lambda)T = 4.6R(P_t + \log[M] - \varkappa)T. \qquad (23)$$

The entropy and enthalpy are determined in this case from the formula

$$\left. \begin{array}{l} \Delta S = -\dfrac{\partial \Psi_\Phi(T)}{\partial T} \cong -\dfrac{\Delta \Psi_\Phi(T)}{\Delta T} \ (\Delta T \text{ small}), \\[2mm] \Psi_H = \Psi_\Phi(T) + T\Delta S. \end{array} \right\}$$

Conclusions

1. When there is a change in the temperature and composition of glasses a fairly significant change is observed in the percentage of cations participating in the electrical conductivity. There is thus clear confirmation, firstly of the existence in glasses of a very low degree of electrolytic dissociation of the polar structural groups, and secondly that the value of the dissociation energy determines the electrical conductivity.

2. The equations for the molar conductivity of borate and silicate glasses in the stable and labile states have been derived theoretically. The particular cases of dissociated cation movement among nonpolar and polar structural elements have been considered.

3. In the stable state the energy parameter of the conductivity $\Psi_\Phi = \Delta\Phi + 2E_a$ does not change with temperature, as there is no change in entropy during electrolytic dissociation.

4. In the labile state the entropy of electrolytic dissociation produces a temperature dependence of the free energy of dissociation.

5. The fundamental experimental and theoretical relationships have been examined, allowing us to produce a quantitative verification of the theoretical expressions so obtained and to determine the entropy of electrolytic dissociation in glasses in the labile state. Formulae are given based on experimental data defining the dissociation energy, Ψ_{Φ}.

References

1. R. L. Myuller (Müller), Phys. Z. Sowjetunion, 1:407, 1932; Ya. I. Frenkel', Z. Phys., 55:652 (1926); Ya. I. Frenkel', A Kinetic Theory of Liquids, Akad. Nauk SSSR, Moscow (1945).
2. R. L. Myuller, Zh. Fiz. Khim., 6:616 (1935); Izv. Akad. Nauk SSSR, Ser. Fiz., 4:607 (1940); ●this volume, p. 170; R. L. Myuller (Müller), Acta Physicochim. URSS, 2:115 (1935).
3. ●R. L. Myuller, Zh. Tekhn. Fiz., 25:236 (1955), ●this volume, p. 15.
4. S. A. Shchukarev and R. L. Myuller, Zh. Fiz. Khim., 1:625 (1930).
5. R. L. Myuller and B. I. Markin, Zh. Fiz. Khim., 5:1272 (1934); R. L. Myuller (Müller) and B. I. Markin, Acta Physiocochim, URSS, 1:160 (1934).
6. B. F. Ormont, Structures of Inorganic Substances, GITTL, Moscow−Leningrad (1950), pp. 178-179.
7. Ya. I. Frenkel', A Kinetic Theory of Liquids, Akad. Nauk SSSR (1945); G. I. Skanavi, Dielectric Physics, GITTL, Moscow (1949).
8. A. A. Lebedev and E. A. Porai-Koshits, Izv. Sekt. Fiz-Khim. Anal. Inst Obshch. Neorgan. Khim. Akad. Nauk SSSR, 16(4):51 (1946); J. Biscoe and B. E. Warren, J. Amer. Chem. Soc., 21:287 (1938); R. L. Green, J. Amer. Ceram. Soc., 25:83 (1942).

III

The Problem of Polarization in an External Field*

It was noted previously that autosolvation, caused by the polarized displacement of univalent undissociated cations in polar structural units, plays an essential part in the electrolytic dissociation of borosilicates [1]. In addition, polarization is also observed under the action of an external field causing, to some extent, this type of displacement and in this way lowering the conductivity. These phenomena must be examined before we begin to compare the experimental quantitative data on electrical conductivity with the theoretical calculations.

1. Electrolysis occurs in borosilicates when a constant field is applied and it is accompanied by electrochemical processes in the sections close to the electrodes. If inert irreversible electrodes are used a considerable concentration polarization appears in their vicinity and this in time sharply decreases the current. The measurements of conductivity under similar conditions do not give the constant reproducible values which characterize a substance of unchanged chemical composition. The latter changes because of the development, near the anode, of a poorly conducting layer depleted in mobile cations. The experimental electrical conductivity obtained under such conditions cannot, of course, be used to test the validity of the theoretical conductivity equation which requires a uniquely defined volume concentration of cations in the dielectric [2]. The values for the conductivity of borosilicates which will be considered later, therefore were obtained from experiments with active reversible anodes, linearly distributed potential gradients, and in the absence of anode polarization [3, 4]. We can also use the results obtained with an alternating field if the glass has a high alkali oxide content and high conductivity.

Even in the absence of anode polarization, a drop in current with time is observed in poorly conducting, low alkali oxide glasses immediately after an external field has been applied.

At high temperatures and with large concentrations of alkali oxide a steady current is established in borosilicates immediately on the application of an external electrical field. In glasses depleted in alkali oxide and at lower temperatures nonstationary, higher initial currents were observed. These drop asymptotically and become steady "residual" currents. The change to the steady state regime proceeds more slowly the lower the concentration of alkali oxides in the glass and the lower the temperature. In the stable region of temperature in the glasses lowest in alkali oxide (in boric anhydride about 200°C) the stationary conductivity generally becomes difficult to measure because of the extreme delay of the drop in the initial current (lasting many tens of minutes). The determination, therefore, of the electrical conductivity of poorly conducting glasses by an alternating current gives the initial excess conductivity which cannot characterize the electrolytic dissociation [5].

When the initial current drops, the distribution of potential remains linear which indicates the onset of volume polarization [4]. The essential role in this volume polarization must be

*R. L. Myuller, Zh. Tekhn. Fiz., 25(9):1556 (1955).

TABLE 1

Compound	Density •	Structural unit	Molecular wt. of structural units	Packing of structural units, mole/ml	Packing of oxygen atoms, mole/ml	Tetrahedral atomic radius, Å	
Boron oxide	1.8	$BO_{3/2}$	34.8	0.052	0.077	B	0.89
Silicon oxide	2.2	$SiO_{4/2}$	60.1	0.037	0.073	Si	1.17

played by the phenomenon of an activated orientated displacement of cations round the ionized atoms of the network structure of the glass. The author has previously considered this in [6] and it is necessary here to consider this phenomenon critically, taking into account the published work from the point of view of the structural chemical features of ionic-covalent borosilicate structures [1].

2. The contemporary physics of amorphous dielectrics distinguishes several types of simultaneously occurring polarization.

An elastic displacement of the valence electron shell with respect to the nucleus occurs in ions and atoms under the influence of an external field. Such polarization is fundamental in pure boron and silicon oxides in which the dielectric constant is only a little larger than the square of the refractive index [7, p. 172]. In boric anhydride the dielectric constant is 3.2 and in fused quartz 3.8. A slight increase is most probably produced by a trace alkali oxide impurity (in boric anhydride it is of the order of 0.01 wt % of sodium oxide [3]).

There is a mistaken idea that in this case the greater packing density of the silicon oxide compared with the boron oxide is important [8]. In fact the reverse is the case. As is shown in Table 1, the structural units in boric oxide are packed 40% more densely than in silicon oxide. The volume concentration of oxygen atoms in boron oxide is also higher (by 5.5%). The higher dielectric constant of silicon oxide may be caused by the greater radius of the valence electron shell of a silicon atom [9] than that of a boron atom, particularly if one considers the 11% decrease in the radius of the latter when it changes to a trigonal structure ($r_{BIII} \approx 0.78$ Å); note the decrease in the distance B—O from 1.53 to 1.36 Å [10]. This effect is weakened by the strengthening of the valence shell as it changes from a six-electron shell in boron and its oxide to the eight-electron shell in silicon (and also its oxide).

These are quite rough calculations. They are, however, clearly closer to fact than the other ideas involving packing densities. Incidentally we may note the lack of any justification for the accepted calculations of the polarizability of boron and silicon in borosilicates by considering them as triple- or quadruple-charged elementary ions which do not in fact exist under normal conditions.

In addition to the elastic electron polarization there is also elastic ionic polarization caused by small oscillatory displacements of cations from their equilibrium position with respect to the ionized atoms of the covalent network. In this case the cation displacements are nonactivated and are not accompanied by their release from the potential wells they occupy.

Polarization due to elastic displacements of electrons and ions occur instantaneously and generally have little dependence on temperature. Therefore, although existing and affecting the internal field in a dielectric, they do not appear to cause any fall in the initial current.

3. As well as these nonactivated cation displacements within the limits of a potential well of a given polar structural element $B^-O_{4/2}M^+$ or $SiO_{3/2}O^-M^+$, activated displacements of cations

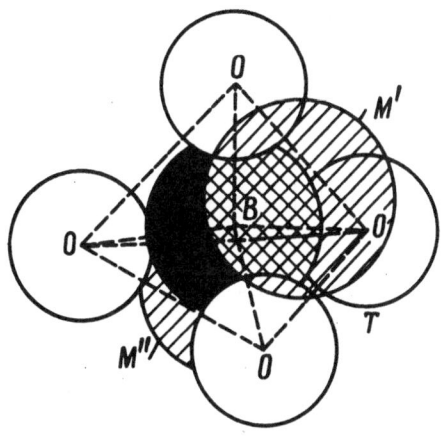
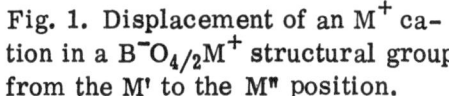
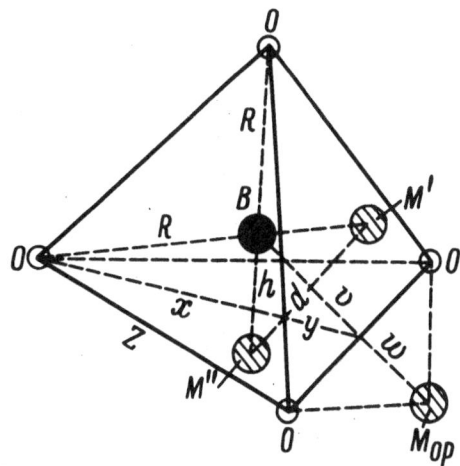

Fig. 1. Displacement of an M^+ cation in a $B^-O_{4/2}M^+$ structural group from the M' to the M" position.

Fig. 2. The geometrical relationship of the M^+ cation displacement in a $B^-O_{4/2}M^+$ structural group.

are possible with a change from one potential well to another in the same structural element. In the borate structural elements there are four such wells, in silicates three [1].

This is illustrated for borates in Fig. 1 and geometrically interpreted in Fig. 2. At the beginning of the displacement the cation bound in the $B^-O_{4/2}M^+$ polar structural group in the position M' falls by the transition T, into the position M". It is not difficult to imagine that the cation, bound electrostatically to the electrically negative boron atom, must be forced during the transition to tear itself away from the boron in order to achieve the cross-over between the oxygen atoms which partly obstruct its path by their valence electron shells. The distance between the cation and boron will in this case be very small compared with the distance on dissociation into the intranetwork space. Accordingly, the activation energy of the orientation shift E_{or} will be much less than the dissociation energy $\Delta\Phi$.

The model of a structural element used here is, of course, only a first approximation to the real picture of the isoelectronic surfaces in oxygen and boron atoms. The present model characterizes the position of the atoms, defines the distance between them, and fixes the four potential wells for the sodium ion around the negatively ionized boron atom. This last property of a structural atomic grouping scarcely undergoes any essential changes when the model in question is subsequently made more precise physically.

Using this model we can continue the attempt to calculate roughly the activation energy of an oriented displacement, starting from the known value of the sum of the boron and oxygen atomic radii in vitreous borates, $R = r_B + r_O = 1.53$ Å [10]. From this figure and in good agreement with the values in [9] of the boron and oxygen tetrahedral atomic radii $r_B = 0.88$ and $r_O = 0.65$ Å, we can make the calculation using Fig. 2. The edge of the tetrahedron, $Z = 2R\sqrt{2/3} = 2.49$ Å. The values of $U = r_O + r_M$ are given in Table 2. In all three cations $U > \frac{Z}{2} = 1.25$ Å, such that the cations cannot pass directly between the oxygen atoms in the plane of the center lines.

The activation energy E_{or} can be obtained in the first approximation as

$$E_{or} = W_{or} - W_{ass} , \qquad (1)$$

where W_{ass} is the energy of the associated M' cation in a potential well; W_{or} is the potential energy of the M_{or} cation in the activated state during displacement. Since

$$W = -\frac{e^2N}{r} = -\frac{329.7}{r} \text{ kcal/mole,} \qquad (2)$$

TABLE 2

Cation	r_M, Å	U, Å	r_{ass}, Å	$-W_{ass}$, kcal/mole	r_{or}, Å	$-W_{or}$, kcal/mole	E_{or}, kcal/mole
Li$^+$	0.78	1.43	1.66	199	(1.57)	199	0
Na$^+$	0.98	1.63	1.86	177	1.92	172	5
K$^+$	1.33	1.98	2.21	150	2.42	137	13

where r is expressed in Å, the activation energy is calculated from the equation

$$E_{or} = 329.7 \left(\frac{1}{r_{ass}} - \frac{1}{r_{or}} \right),$$

where r_{ass} and r_{or} are the distances in Å between the centers of the boron atom and the cation in a potential well and in an activated state, respectively; here $r_{ass} = r_B + r_M$. In fact, from elementary geometry it is easy to check that in these particular conditions the cations in the potential well will "lie on" the boron atom, touching at the same time not more than two oxygen atoms.

From Fig. 2, $r_{or} = v + w$ when $v = \sqrt{h^2 + y^2} = R/\sqrt{3}$ such that $h = R/3$ and $y = R\sqrt{2}/3$; $w = \sqrt{u^2 - Z^2/4}$.

It is clear from Table 2 that $r_{or} > r_{ass}$ in oriented displacements of sodium and potassium ions; these changes are thus activated. In the case of lithium, r_{or} is less than r_{ass}, which indicates that an unimpeded glide of this cation along the boron atom can occur during this particular displacement (above the center lines of the oxygen atoms). From the values of the atomic radii above we should expect an absence of activated polarization displacements in lithium borates.

The activation energy E_{or}, calculated from Eq. (3), of the orientational displacements of sodium and potassium ions and the W_{ass} and W_{or} parameters are given in Table 2. The value of E_{or} is small but larger for potassium than for sodium. For lithium, the activation energy obtained is zero and therefore there is an activation energy series

$$E_{or}(Li) < E_{or}(Na) < E_{or}(K). \tag{4}$$

The interaction of the dipolar structural element with the surrounding medium was not taken into account in calculating these parameters. A relatively slight increase in this interaction can be assumed if there were an insignificant rise in the dipole moment (<10%) on cation activation.

In the case of oriented cation displacement in the silicate $SiO_{3/2}O^-M^+$ structural group with three potential wells round the ionized oxygen atom, the activation may result from the breakdown of the polarization bond with the silicon atom. However, if one considers the value of the atomic radius of silicon to be 1.17 Å [9], then for all three cations in question one finds $r_{ass} > r_{or}$, i.e., $E_{or} = 0$.

But, as was indicated above, such calculations cannot claim to be strict. The natural valence chemical bonds to boron and silicon atoms have an ionic–covalent character and thus the atomic radii of boron and silicon must be a little lower and the atomic radius of oxygen higher than the values taken above. We should, therefore, expect also to find an activated displacement in alkali metal silicates and in lithium borates. The activation energies must then be a little higher than the values shown in Table 2.

This mechanism alone of activated ionic polarization can to some extent explain the observed fall in the primary current in borosilicates when an active reversible anode is used. In this case the decrease in the rate of establishing a stationary current with a decrease in temperature and the effect of stabilization of the glass may be explained by the freezing of the valence degrees of freedom of the vibrational motion of the boron and oxygen atoms as stable bonds are formed between them. On average this must slightly increase the height of the potential barrier of the cation displacement inside the polar structural groups.

4. Orientation displacements of cations around the immobile ionized boron and oxygen atoms differ in their sturctural chemistry and kinematics from the rotational motion of polar molecules in liquids. In the resulting polarization, however, there is no difference between the two types of dipole orientation in principle.

Aleksandrov, Kobeko, and Kubshinskii studied the appreciable losses in borosilicates which were not explained by conduction loss and were analogous to those caused by the rotation of polar molecules [11]. It was not possible to explain the increased polarization by a rotation of molecules since molecules are absent in borosilicates and in the solid state they would be unable to rotate.

It is now possible to suggest that the mechanism of dipole orientation is the displacement of the univalent cations in the potential wells of the structural covalent units. It follows that ion-covalent structural units must be present for this increase in polarization. Therefore a fundamental difference between borosilicates and purely ionic crystalline substances in which no increased high frequency losses are observed, become clear. The presence of increased polarization and high dielectric losses in borosilicates with univalent cation orientation and the absence of such phenomena in borosilicates with orientationally nondisplaceable divalent cations, are fundamental [1]. In fact in the last case, small dielectric losses and constants are observed with a considerably lower temperature dependence than that of alkali glasses.

5. This mechanism of activated ionic polarization in borosilicates arises from the chemical structural features of the latter and explains fairly concretely and logically the known facts. In this respect the vague suggestions presented by Skanavi based on a "thermal ionic polarization" hypothesis must be noted. Starting from the soundly based and correctly evaluated experimental facts presented by Aleksandrov, Kobeko, Kubshinskii, et al. Skanavi put forward slipshod suggestions unsound from the point of view of atomic structured substances.

According to the thermal ionic polarization hypothesis, in glass the metal ions are present in regions which are loosely packed. These metallic ions can complete a "significant movement inside this weak packing" without participating in the conductivity [12]. These movements occur in "particular, limited regions" [7, p. 419] "over a distance comparable with the size of the molecules" [7, p. 41] and also "over significant distances, perhaps up to 10 atomic distances" [12]. The absence of the structural chemical basis necessary for a concrete definition of the material features of the process is obvious here. Moreover, this thermal ionic polarization hypothesis is the basis for the mistaken idea that in borosilicates, cation dissociation is complete and their coherence relatively weak.

As well as the thermal ionic polarization hypothesis, Skanavi discussed in a mathematically valid way the kinetics of the polarization process in terms of a cation flipping from one potential well into the neighboring well over a low potential barrier when both potential wells are bounded externally by high potential barriers [7]. Thus, the mathematical section of the hypothesis may be directly attributed to the activated displacement of cations inside the polar structural elements.

According to the above, in a polar structural group containing boron the associated cation may occupy any one of four potential wells. The location of a cation in one or the other potential well is statistically random and is due to the disordered thermal activated oscillations of the cation. When an external electrical field is applied, a small proportion only of the cations is displaced in the direction of the field because of the thermal disorder at a given temperature.

In a disordered network the tetrahedral structural groups have a random orientation and on average there will be a pair of potential wells at each element arranged in succession along the direction of the external field. Starting from a condition in which initially one and a half cations are located in the "lefthand" wells and the other one and half in the right, and by introducing minor changes in the Skanavi formula, it is possible according to Skanavi to obtain an expression for the number of cations per ml displaced in the structural elements in the direction of the field at a time t [7, p. 43].

$$\Delta n = \frac{n\delta FH}{4RT}\left(1 - e^{-\frac{t}{\tau}}\right), \\ \tau = \frac{3\exp(E_{or}/RT)}{\nu},$$ (5)

where n is the number of polar structural elements per ml; δ is the average shift in the direction of the field; F is the Faraday constant; H is the field strength; R the gas constant; T the absolute temperature; ν is the frequency of the thermal oscillations of the cations, approximately 10^{13} sec^{-1}; and τ is the relaxation time.

According to Skanavi's results the value of the activation energy is about 10^{-12} erg [7, p. 323], or 14 kcal/mole, close, in order of magnitude, to the calculations shown in Table 2. Such a low activation energy is responsible for the small value of τ which even at room temperature (300°K) is, according to [5], only 0.003 sec.

It has already been seen that a slowly increasing (over tens of minutes) volume polarization is characteristic even at 200°C of glasses low in alkali oxides. It would seem from this that, as well as those discussed above, there must be one type of polarization at least with a significantly greater energy, commensurate with the dissociation energy of a cation (according to [5], when t ≈ 600 sec. T = 500°K, E_{or} = 35 kcal/mole).

6. In borosilicates with a low alkali oxide content, there is typically the development of groups of associated polar structural groups. These are isolated by the basic matrix of nonpolar structural units [6]. Conductivity within the polar groups must in this case be higher than the conductivity of the surrounding nonpolar medium. When an external field is applied, at the boundary transition layer between the polar and the nonpolar matrix concentration polarization phenomena can appear and so produce macroscopic volume polarization. The rate at which this polarization is established will be determined by the conductivity and consequently by the appreciable dissociation energy [2]. A similar polarization is possible in borosilicates in which there are differentiated groupings in the nonpolar medium, but such polarization is impossible in purely ionic crystals because of the absence of similar structural features. In the presence of divalent cations and the absence of any alkalis, particularly sodium, this phenomenon would vanish in borosilicates because of the poor conductivity of the relevant polar structural groups.

Thus, as well as the activated polarization caused by the oriented displacements of cations within the polar structural units with an activation energy of the order of 20 kcal/mole or less, activated polarization of borosilicates is also possible as a result of their nonhomogeneous chemical composition. This polarization is possibly the basic cause of the observed fall in primary current.

7. In the particular group of dielectrics which belong to the ionic-covalent systems, anomalously high polarization and dielectric losses related to those in molecular ferroelectrics is observed in the intermediate temperature region [7]. A series of physical studies indicated that molecular ferroelectrics may exhibit polar molecular orientation of radicals or complexes in an external field. Such an orientation is absent at low temperatures due to the freezing of the torsional vibration degrees of freedom and absent at sufficiently high temperatures as it must

be suppressed by the intense thermal vibrations of the particles. This phenomenon is complex in character and depends on fine structural chemical features of the substance. To date this has still not been adequately studied.

A reason for the anomalously high polarization of some ionic-covalent dielectrics may also be found in their structural chemical features. As is known, rutile, titanates and some other substances of a specialized composition and structure, are related to these dielectrics [7].

It has been suggested that impurity ions, for example barium in $BaTio_3$, play the chief role in the development of ferroelectric properties. It is proposed that the free barium ions, displaced from the regular lattice by other impurity oxides, are uniformly distributed in the interstices deforming the neighboring lattice cells and forming regions of "spontaneous" polarization. This supposition does not take into account the ionized oxygen atoms which also appear together with the impurity ions. The former are most probably connected to the basic covalent network and fix the barium ions by developing associated polar groups analogous to those in borosilicates. The distribution of the excess "impurity" ions in barium titanate can, therefore, hardly be uniform.

According to Skanavi's work [7,12] the high dielectric constant of rutile and of titanates with the perovskite structure is due to an ionic displacement specific to these structures, in the internal Lorentz sphere, which sharply increases the dielectric constant. At the same time it is suggested that an increased electron polarization is present, caused by the significant polarizing tendency of the ionized oxygen atoms. The presence in the "spontaneous polarization" regions of weakly bonded electrons, easily polarized in a field, is also noted in [13]. The essential role of chemical impurities was subsequently established; when these are not present the anomalous phenomena vanish. The impurities probably produce a special type of tetragonal structure very similar to that of perovskite [7, p. 192]. In particular, Skanavi suggested that, in addition to polarization resulting from electron and ion displacement, titanates with a perovskite structure must have specifically increased polarization produced by the "polarity of individual groups in the crystalline lattice" [12].

8. In the light of this brief survey of the facts and views concerning the anomalously high polarizing tendencies and dielectric losses in groups of ion-covalent substances with the perovskite and rutile structures, we should consider some of the kinetic features of these structures derived from thermal capacity data [6].

It is known to be unlikely that elementary ions with high charges exist in solids under normal conditions because of their enhanced interaction. We should, therefore, naturally expect that thermal vibrations would make possible some association of such ions in the crystalline lattice, however short-lived, with the formation of covalent-like bonds between them. This does not exclude the possibility that ions such as Ti^{4+} and O^{2-}, on a statistical time-average form complex ions with decreased effective charges or at least with charges dispersed over some of the atoms in the complex, for example $[TiO]^{2+}$ and $[TiO_3]^{2-}$.

The latter structural formulae express the time-average of the statistical effect of the association of the ions, analogous to the Debye-Hueckel ionic clouds in solutions. The thermal capacity of TiO_2 shows that at normal temperatures rigid valence bonds, consistent with the formulae for the complexes, exist between the ionized titanium and oxygen atoms (four bonds in $2TiO_2$ [6].

The fluctuation of the valence bonds is accompanied by changes in the atomic partners. The frequency of such fluctuation must lie within the limits $1 \ll \nu \ll 10^{13}$ sec^{-1}. In fact the relaxation time must be small since it is not detected in the x-ray structural analysis of such complexes in rutile and perovskite. Moreover, the lifetime of such rigidly bonded complexes must be fairly long compared with the thermal vibrational period of the atoms, since such rigid bonds

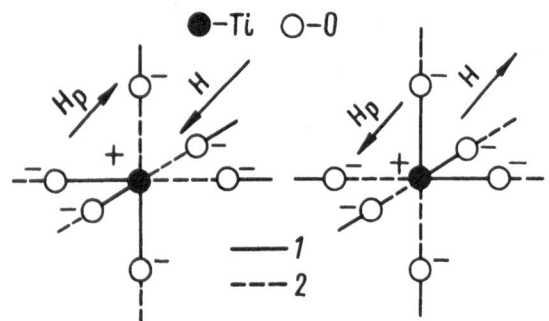

Fig. 3. The ionic—covalent bonds in a $[TiO_{6/2}]^{2-}$ structural group of perovskite when the direction of the external field H is changed. 1) Rigid valence bond; 2) ionic bond.

are reflected in the lowering of the thermal capacity. A fluctuation, as temporary structures arise, is possible when, in particular, there is a structure with a number of coordinating atoms which exceed the number of simultaneously existing rigid bonds. In fact, from the x-ray analysis of both rutile and perovskite there are six coordinating oxygen atoms in the lattice and only one or three simultaneously realized rigid covalent bonds:

$$n\,[TiO_{6/3}] \rightarrow \frac{n}{2}\left[\overrightarrow{TiO}\right]^{2+} + \frac{n}{2}\left[\overrightarrow{TiO_3}\right]^{2-},$$
$$n\,[TiO_{6/2}]^{2-} \rightarrow n\left[\overrightarrow{TiO_3}\right]^{2-}. \qquad (6)$$

It is not difficult to check from the crystal structure of rutile and perovskite that all ionized complexes must be unsymmetrical and must have considerable dipole moments. In the absence of an external field, fluctuating dipoles have statistically at any given moment a completely disordered orientation. When an external field is applied, however, the polar groups become partly oriented in the direction of the field H (Fig. 3).

The relaxation time of such polarization, $\tau = 10^{-8} - 10^{-9}$ sec, found by Skanavi [7, p. 436] is entirely consistent with the limits for the fluctuation frequency of the polar complex discussed above.

The well known weakening of this effect of an external field will be observed when there is a decrease in the radius of the "impurity ions", as a result of the increased polarizing effect of the latter on the polar complexes [12]. The decrease in the impurity-ion effect is noted in zirconates and stannates with a perovskite structure when there is an increase in the divalent cation radius [4].

Dielectric losses must decrease above a particular temperature, in view of the disappearance of the rigid valence bonds when the valence oscillations are excited with the consequent break-up of the polar complexes.

At low temperatures we should also expect a lower loss since, in this case, the restructuring fluctuations of the complexes must become more difficult as a result of the decrease in the intensity of the thermal vibrations.

It is very important to check these ideas by x-ray analysis of the structures of the substances in question in a sufficiently strong electrical field to ensure an adequate percentage of oriented polar complexes.

We must note, in conclusion, that in some measure or other all ionic-covalent substances must have the property of developing short-lived oriented dipole ionic-atomic complexes. In all probability the so-called "structural losses" at low temperatures and high frequencies are connected with this. It is a mistake to attribute the low temperature losses in silicate glasses to deformation of the vitreous network [15]. A vitreous network cannot be deformed at low temperature owing to the absence of the valence degrees of freedom of the atomic vibrations [6, 16].

Conclusions

1. The polarization of ion-covalent materials is characterized by the following features which are absent in purely ionic crystalline materials:

a. together with the instantaneously occurring elastic displacements in the potential wells of bonded ions, there is activated ionic polarization involving the same bonded cations as a result of their transition from one potential well to another of the same structural group:

b. together with the instantaneous elastic electron displacement in the atoms there is an orientation of polar complexes which occurs at a frequency $1 \ll \nu \ll 10^{13}$ sec^{-1} as a result of the temporal fluctuations in the distribution of valence electrons of the structural atomic groups.

2. Activated ionic polarization explains the increased dielectric loss at low temperatures in borosilicates and related substances low in alkali oxides.

3. The orientation of the fluctuating polar ionic-covalent complexes is the most likely explanation for the anomalously high polarizability and dielectric loss in substances with the perovskite or rutile structure.

4. The fluctuation of oriented polar complexes in ion-covalent systems is possible in certain structural conditions, in particular, with a fairly large number of coordinating atoms and with a weak polarizing effect of "impurity ions." This phenomenon requires the presence of an interaction between high valence elements where the product of their valences is not less than six.

5. The existence of a critical region of anomalously high dielectric loss is due to the vanishing of the rigid valence bonds at high temperatures and to the inhibition of the fluctuations in oriented polar complexes at low temperatures, consequent on the decrease in intensity of the thermal valence vibrations.

6. The decrease in the primary current in low alkali oxide depleted borosilicates in the absence of electrode polarization may be due to both activated ionic polarization and to the polarization of the better conducting associated polar inclusions in the basic nonpolar medium.

7. The mechanism of activated ionic polarization was considered and the value of the corresponding activation energy was approximately determined:

$$E_{or}(Li) < E_{or}(Na) < E_{or}(K).$$

8. The related features in molecular and ionic-covalent orientational polarization were noted. Criticisms were expressed of the thermal ionic polarization hypothesis and of the suggestion that low temperature losses can be attributed to a deformation in the vitreous network.

References

1. R. L. Myuller, Zh. Tekhn. Fiz., 25:236 (1955), ● this volume, p. 15.
2. R. L. Myuller, Zh. Tekhn. Fiz., 25:246 (1955), ● this volume, p. 24.
3. S. A. Shchukarev and R. L. Myuller, Zh. Fiz. Khim., 1:625 (1930).
4. R. L. Myuller and B. I. Markin, Zh. Fiz. Khim., 5:1272 (1934).
5. R. L. Myuller and B. I. Markin, Zh. Fiz. Khim., 5:1262 (1934); R. L. Myuller, Uch. Zap. Leningr. Gos. Univ., No. 54, 159 (1940), ● this volume, p. 3.
6. R. L. Myuller, The Vitreous State and the Electrochemistry of Glass (Doctoral Dissertation). Leningr. Gos. Univ. (1940).
7. G. I. Skanavi, Dielectric Physics, GITTL, Moscow. (1950).
8. G. I. Skanavi, Zh. Tekhn. Fiz., 7:1039 (1937).

9. B. F. Ormont, Structures of Inorganic Substances, GITTL, Moscow (1950), pp. 178–179.

10. A. A. Lebedev and E. A. Porai-Koshitz, Izv. Sekt. Fiz. Khim. Anal. Inst. Obshch. Neorgan. Khim., Akad. Nauk SSSR, 16 (4):51 (1946).

11. A. P. Aleksandrov, P. P. Kobeko, and E. V. Kuvshinskii, Zh. Tekhn. Fiz., 6:963 (1936).

12. G. I. Skanavi, Élektrichestvo, No. 8, p. 15 (1947).

13. E. V. Sinyakov, E. A. Stafiichuk, and B. K. Chernykh, Zh. Éksp. Teor. Fiz., 21:610 (1952).

14. G. A. Smolenskii, Zh. Tekhn. Fiz., 22:3 (1952).

15. V. A. Ioffe, "Dielectric losses in silicate glasses," in: The Structure of Glass, Vol. 1, Consultants Bureau, New York (1955), p.

16. R. L. Myuller, Zh. Fiz. Khim., 28:1193 (1954).

IV

The Problem of Polarization and Electrolytic Phenomena*

The Electroconductivity of a Material Polarized
by an External Field †

1. In the preceding paper it was emphasized that the specific electrical conductivity can only be measured and used uniquely to characterize a substance of a particular chemical composition and, above all, of a particular volume concentration of cations, when there is no concentration polarization in the region close to the electrodes [1]. It has been known for some time that this polarization can be eliminated by using reversible active anodes [2]. Only such true conductivity will be considered here.

In the absence of the surface effect of electrode polarization, however, there is a decrease with time in the initial current as a result of the development of volume polarization. An extensive series of investigations, many by Soviet physicists, has been devoted to this problem [3].

From the abundant experimental results and the additional information relating to polarization in an external field in [1], we can describe the physical nature of the steady-state conductivity in the following way.

Immediately after the application of an external field H the specific conductivity is observed to be

$$x_0 = \frac{i_x + i_A + i_\mu}{H - H_p},$$ (1)

where i_x is the current density of the true conduction current. In this case the strength of the external field H will be somewhat lowered by the counterfield H_p created by the elastic ion and electron polarization. The initially increasing activated ionic polarization H_A and the polarization of the associated polar microinclusions H_μ in the field direction, although perhaps very small nevertheless cause noticeable polarization displacement currents i_A and i_μ. This nonstationary specific conductivity is poorly reproducible and cannot be used as a characteristic parameter for a substance of a specified chemical composition.

*R. L. Myuller, Zh. Tekhn. Fiz., 25(9):1567 (1955).
†It should be remembered that there is another approach, different from that presented here, widely used to describe polarization phenomena. The strength of the external field is given by the voltage of the source connected to the electrodes while the increase in the polarization of the specimen increases the charge at the electrodes but cannot decrease the field strength and consequently $H_\infty = H$. For a discussion of this problem see the selection, Electrical Properties and the Structure of Glass, Izd. Khimiya, Moscow−Leningrad (1964); p. 93. (Editor's Note)

2. The steady state stationary ("residual") specific conductivity in a reproducible physical parameter

$$\varkappa_\infty = \frac{l_\varkappa}{H_\infty},$$ (2)

which is observed after the complete development of the activated ionic polarization H_A and the polarization of associated polar microinclusions H_μ. Under these conditions evidently there are the equations

$$i_A = i_\mu = 0,$$
$$H_\infty = H - (H_p + H_A + H_\mu).$$

It is as yet impossible, even in the case of alkali halide crystals, to calculate theoretically the dielectric constant because of the difficulty of determining the internal field in a crystal [4]. It is even more difficult to calculate the field inside the Lorentz sphere in nonuniform vitreous covalent-ionic systems.

Because it is not possible to determine the value of the actual internal field strength H_∞ the determination of the experimental steady state specific conductivity from the relationship

$$\varkappa_e = \frac{l_\varkappa}{H}$$ (3)

is generally adopted using the external field H without allowing for the counter emf which arises as a result of the steady state volume polarization inside the specimen under investigation.

The strict theoretical expression for specific electrical conductivity is of course Eq. (2) which involves the true internal field strength in the specimen [5]. Thus, between the experimental values $\varkappa_e$ and the theoretical $\varkappa_t$ for specific conductivity there is the relationship

$$\varkappa_e = \varkappa_t \frac{H_\infty}{H} = \varkappa_t (1 - \beta),$$ (4)

where

$$\beta = \frac{H_p + H_A + H_\mu}{H} < 1.$$

After transforming both parts of Eq. (4) to volume molar concentration of cations we obtain an analogous relationship for molar conductivity, theoretical Λ_t and experimental Λ_e,

$$\Lambda_e = \Lambda_t (1 - \beta).$$ (5)

By expressing the experimental and theoretical molar conductivity in the form [5, Eq. (18)]

$$\Lambda = 10^p \exp\left(-\frac{\Psi_\phi}{2RT}\right)$$

and introducing the factor $(1-\beta)$ into the pre-exponential term we obtain, in the logarithmic form,

$$p_e = p_t + \log(1 - \beta).$$ (6)

3. Over a more or less narrow temperature range, in the region of 20–300°C in alkali glasses, the decrease in the primary current with time is no longer observed. This is not due to the thermal dissipation of volume polarization charges since the relative changes in the absolute temperature are insignificant. Nor is this cessation of the observed decrease in current due to an increase in the rate of completion of the polarization process.

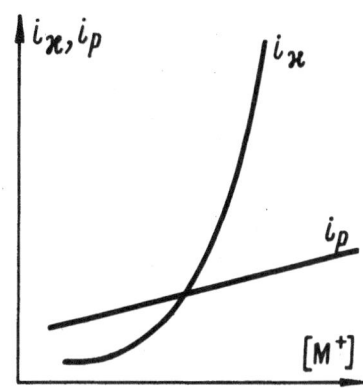

Fig. 1. The dependence of the continuous conductivity current density $i_\varkappa$ and of the activated polarization current density i_p on the volume concentration of polar structural units which is equal to the concentration of M^+ cations.

As Venderovich's experimental studies showed [3] we are dealing here essentially with the disappearance of volume polarization. It is most likely that polarization ceases to be observed due to the masking of the polarization current i_p by the larger continuous conductivity current $i_\varkappa$. The high temperature coefficient of the latter is caused by its appreciable activation energy which exceeds the activation energy of activated polarization. The decrease in the polarization of glasses with an increase in the alkali cations concentration, established experimentally by Venderovich, is in good agreement with this view.

The initial activated polarization absorption current i_p increases roughly linearly with the volume concentration of polar structural units while the continuous conductivity current $i_\varkappa$ increases exponentially. Thus, with an increase in the cation concentration in borosilicates the continuous current overtakes and considerably exceeds the initial polarization current (see Fig. 1).

The cross-over occurs, of course, at a higher cation concentration and at a higher temperature if the cation dissociation energy is larger, for example in glasses which contain dielectrovalent barium cations. The experimental results of Venderovich for barium borate glasses confirm this.

The use of graphite electrodes contacted by a metal block (developing a poorly conducting layer close to the anode) can explain the marked polarization observed by Venderovich in some glasses rich in alkali cations. The effect on the polarization of the substitution of silica by boric oxide was not strictly examined by Venderovich: he did not measure the change in the boron, silicon and sodium content in molar volume concentration units. The use of such units exposes more clearly the effect of the concentration factor on the electrochemical parameters.

According to Venderovich's results for glasses depleted in alkali cations [3], the polarization potential at low temperatures (like the polarization field strength H_∞) reaches 90% of the value of the external voltage ($\beta \simeq 0.9$). Thus $\log (1-\beta)$ can be reduced to -1. At higher temperatures $\log (1-\beta)$ approaches zero (the polarization voltage is small compared with the external voltage).

Except for the narrow interval of temperature indicated above, which corresponds to the rapid increase in the value of the parameter β, the temperature coefficient of the polarization voltage is insignificant.

Turning to Eq.(6) we can now state that the theoretical value of the statistical power-parameter p_t to an accuracy (± 1) retains its value even in the presence of a polarization field in the glass.

The Local Polarization of a Medium by Dissociated Cations and the Free Energy of Electrolytic Dissociation

1. Consideration has for a long time been given to the fact that the electrical conductivity of borosilicates with a low alkali oxide content must be limited by the movement of the weakly dissociated cations in a nonpolar silica-boric oxide medium and at high alkali concentration by the more highly dissociated cations and vacant sites in a polar medium [6]. The superficial

similarity in the concentration dependence of electrical conductivity in borosilicates and in liquid solutions of electrolytes in nonpolar solvents was also noted.

The essential role in increasing electrolytic dissociation was assigned to the phenomena of the association of polar inclusions and the solvation of dissociated cations by polar groups of the electrolyte itself (autosolvation) [7]. Frenkel' later suggested an approximate calculation for the decreased dissociation energy of the polar molecules as a result of the action of the surrounding polar molecular medium treated as a continuum with a specified dielectric constant [8].

In the author's earlier papers and also in Frenkel's, the analysis of the electrolytic dissociation in glasses was based on the assumption of the molecular-dispersed structure of borosilicates. After the significance of the covalent structural form of matter in the vitreous state had been established, the author in due time introduced an essential modification: the concept of electrolytic dissociation accompanied by a solvation effect caused by oriented displacements of the cations bonded in polar structural elements [9].

Assuming that glass has basically two forms of structural units, nonpolar (such as $BO_{3/2}$) and polar ($B^-O_{4/2}M^+$), we can extend the approximate method of Frenkel' to calculate the solvation effect. This effect causes an abrupt change in the free energy of electrolytic dissociation in the changeover from molar conductivity in a nonpolar medium (low alkali oxide concentrations) to conductivity in a medium of polar structural groups (high alkali oxide concentrations).

We have already stated that it is not yet possible to make a strictly valid calculation of the internal field in a dielectric without having to introduce well known simplifications. The results, therefore, presented below, should be taken only as semiquantitative basic aids to the correct understanding of the nature of the exponential growth of electrolytic dissociation in borosilicates with an increase in the polar structural group concentration.

2. There are two distinct types of electrolytic dissociation in borosilicates. When the dissociated ions are in nonpolar or polar media we have, particularly in the case of borates of univalent metals, two equations [5]:

$$B^-O_{4/2}M^+ + BO_{3/2} \rightarrow B^-O_{4/2} + M^+BO_{3/2}, \tag{7}$$

$$B^-O_{4/2}M^+ + B^-O_{4/2}M^+ \rightarrow B^-O_{4/2} + M^+B^-O_{4/2}M^+. \tag{8}$$

These equations are respectively satisfied by the two expressions for the change in free energy (for glass in the stable state):

$$\Delta\Phi = -(W_{ass} + W_p) + (W_+ + W_-), \tag{9}$$

$$\Delta\Phi' = -(W_{ass} + W_p) + (W'_+ + W_-). \tag{10}$$

Here

$$W_{ass} = -\frac{e^2N}{r} = -\frac{330}{r^0_{ass}} \text{ kcal/mole} \tag{11}$$

expresses the potential energy of a cation associated with a negatively ionized boron atom in a $B^-O_{4/2}M^+$ structural element; $r^0_{ass} = r^0_B + r^0_M$ is the sum of the radius of the boron and the cation, in Å.

The potential energy of the interaction between the polar structural group element ($B^-O_{4/2}M^+$) and the surrounding medium with a dielectric constant ε is approximately given, using the Frenkel'-Onsager concepts [8,10] as;

$$W_p = -\frac{\varepsilon - 1}{2\varepsilon + 1} \cdot \frac{p^2N}{R^3} = -\frac{N(\varepsilon - 1)e^2r^2_{ass}}{(2\varepsilon + 1)R^3} = -\frac{330(\varepsilon - 1)}{(2\varepsilon + 1)} \cdot \frac{r^{02}_{ass}}{R^{03}} \text{ kcal/mole.}^* \tag{12}$$

*In this paper, Myuller uses r and R to denote the radii in arbitrary units and r^0 and R^0 to denote the radii in angstroms (Editor's Note).

where $p = er_{ass}$ is the dipole moment of a structural group and $R^0 = r_B^0 + r_O^0 = 1.53$ Å is its radius [1].

The potential energy of interaction between the dielectric medium and the vacant negatively ionized $B^-O_{4/2}$ structural group is determined from

$$W_- = -\frac{e^2 N}{2a}\left(1 - \frac{1}{\varepsilon}\right) = -\frac{330}{2R^0}\left(1 - \frac{1}{\varepsilon}\right) \text{kcal/mole,} \tag{13}$$

where $R^0 = 1.53$ Å is the constant radius of the tetrahedral structural unit [1].

The potential energy of the interaction between the dielectric medium and a dissociated cation linked to a nonpolar $M^+BO_{3/2}$ structural unit is

$$W_+ = -\frac{330}{2r^0}\left(1 - \frac{1}{\varepsilon}\right) \text{kcal/mole,} \tag{14}$$

where $r^0 = r_{BIII}^0 + r_O^0 = 1.36$ Å is the radius of the trigonal nonpolar structural element [1].

Finally, the potential energy of a dissociated cation bonded in a $M^+B^-O_{4/2}M^+$ polar structural unit and interacting with the surrounding polar medium is expressed in the form

$$W'_+ = -330\left[\frac{1}{r_{ass}^0} - \frac{1}{d^0} + \frac{1}{2R^0}\left(1 - \frac{1}{\varepsilon}\right)\right] \text{kcal/mole.} \tag{15}$$

The first term allows for the attraction between the dissociated cation and the negative boron; the second term is the mutual repulsion of two cations; the third term is the interaction of the surrounding medium with an effective positively charged $M^+B^-O_{4/2}M^+$ polar structural group with radius $R^0 = 1.53$ Å. The dipole interactions of Eq. (12) are superimposed on this; to a first approximation these remain unchanged and are therefore eliminated from the calculation. From the geometry of Fig. 2 shown in [1] it follows that the distance between the centers of the cations $M'M''$ may be closely described by

$$d = \frac{Z}{R} r_{ass} = 1.63 r_{ass}^0 \text{ Å,}$$

where $Z = 2.49$ Å is the edge of the tetrahedron, and $R = 1.53$ Å.

In all these calculations we have not considered the relatively much smaller value of the interaction between the electron shell of the cation and the electron shells of the nearby atoms, nor the particularly active polarization interaction with the latter.

In the case of cations with a valency z the values of $\Delta\Phi$ in Eqs. (9) and (10) must be multiplied by z.

3. Investigations by Kobeko, Skanavi, Bogoroditskii, et al. have shown that the dielectric constant increases from 3 to 16 on the transition from a simple quartz glass to an 80% lead silicate glass. An analogous rise is observed in other glasses as the concentration of their polar component is increased [11].

We shall now consider the calculated values of the free energy of electrolytic dissociation $\Delta\Phi$ in alkali borates for two cases: 1) dissociation according to Eq. (7), with a low alkali oxide content, and a dielectric constant $\varepsilon = 6$; 2) dissociation according to Eq. (8), with a considerable alkali oxide content and a dielectric constant $\varepsilon = 18$.

The calculations were made using Eqs. (9) and (10) and the formulae of Eqs. (11) to (15). Table 1 shows the results.

The calculated results for the free energy so obtained, although quite approximate, nonetheless give a correct picture of the increase in the autosolvation effect as the concentration of

TABLE 1

M	r_M^0	r_{ass}^0	d^0	W_{ass}	$\varepsilon = 6$				$\varepsilon = 18$			
					$-W_p$	$-W_+$	$-W_-$	$\Delta\Phi$	$-W_p$	$-W_+'$	$-W_-$	$\Delta\Phi'$
	Å			kcal/mole (cation)								
Li	0.78	1.66	2.71	199	102	101	90	110	116	178	102	35
Na	0.98	1.86	3.04	177	128	101	90	114	145	172	102	48
K	1.33	2.21	3.60	150	180	101	90	139	205	158	102	95

polar structural groups is increased producing an increase in the dielectric constant. As Frenkel' noted, the dissociation energy $\Delta\Phi$ must fall considerably in this case and there is therefore, according to the equation for electrical conductivity, an exponential increase in the latter [6, 7]. At the same time an increase is noted in the energy $\Delta\Phi$ through the series

$$\Delta\Phi(\text{Li}) < \Delta\Phi(\text{Na}) < \Delta\Phi(\text{K}), \qquad (16)$$

which produces the experimentally observed [12] increase in mobility in the series of borates

$$\Lambda(\text{Li}) > \Lambda(\text{Na}) > \Lambda(\text{K}). \qquad (17)$$

In the case of dissociation involving the discharge of cations from associated polar micro-inclusions into a nonpolar medium we should further consider the Madelung coefficient α_M not taken into account in Eq. (11), and so

$$W_{ass}' = -\frac{330}{r_{ass}^0}\alpha_M \text{ kcal/mole.} \qquad (11a)$$

The Madelung coefficient must increase with an increase in dipolar interaction, i.e. from lithium to potassium, and will thus produce according to Eq. (9) a corresponding increase in the dissociation energy. In this case the steps in the free energy and mobility between the members of the series of Eqs. (16) and (17) will increase.

Conclusions

1. The parameter defining specific electrical conductivity with the most physical validity and which uniquely characterizes borosilicates of a given chemical composition (and thermal history), is the stationary ("steady-state," "residual") specific conductivity. This is determined experimentally by using reversible active electrodes which ensure the absence of concentration polarization in the region close to the electrode.

2. Nonstationary electrical conductivity, including ac conductivity, is not in many cases a parameter which can uniquely characterize a substance of a given chemical composition.

3. A correction for the lowering of the external field by polarization phenomena slightly changes the theoretical value of the power parameter P_t in the molar conductivity expression

$$P_t^{\text{corr}} = P_t^{\text{calc}} \pm (1-2).$$

4. The polarization of an ion-covalent medium by polar structural groups and by the products of their electrolytic dissociation involves the oriented displacements of undissociated cations in surrounding polar structural groups (autosolvation).

5. The final approximate calculation of the free energy of electrolytic dissociation in vitreous borates of alkali metals in nonpolar and polar media indicates that the autosolvation effect should reduce the free energy of electrolytic dissociation.

6. The mobility series in borate glasses

$$\Lambda(Li) > \Lambda(Na) > \Lambda(K)$$

has a theoretical basis.

References

1. R. L. Myuller, Zh. Tekhn. Fiz., 25:1556 (1955), ● this volume, p. 33.
2. H. Buff, Ann. Chem. Pharm. (Lieb.), 90:257 (1854); W. Beetz, Ann. Phys. Chem. (Pogg.), 92:462 (1954); E. Warburg, Ann. Phys. (Wied.), 21:622 (1884); M. LeBlanc and F. Kirschbaum, Z. Phys. Chem., 72:168 (1910); S. A. Shchukarev and P. P. Kobeko, Proc. of the Chemical Section, Russian Phys. Chem. Soc., 32(7):10 (1926).
3. A. A. Shaposhnikov, Zh. RFKhO Otd. Fiz. (Journal of the Russian Physical Chemical Soc., Physics Section), 42:376 (1910); A. F. Ioffe, Izv. Petrograd. Politekhn. Inst. Otd. Tekhn., Estestvosn., mat., 24(1):1 (1915); Crystal Physics, Gosizdat, Moscow—Leningrad (1929); A. F. Val'ter (Ed.), Dielectric Physics, Gosizdat, Moscow—Leningrad (1933); G. I. Skanavi (Ed.), Dielectric Physics, GITTL, Moscow—Leningrad (1949); A. M. Venderovich, Zh. Tekh. Fiz., 23:282 (1953).
4. N. Mott and R. Gurney, Electron Processes in Ionic Crystals, Oxford Clarendon Press (1940).
5. R. L. Myuller, Zh. Tekhn. Fiz., 25:246 (1955), ● this volume, p. 24.
6. R. L. Myuller (Müller), Phys. Z. Sowjetunion, 1:407 (1932); Nature, 129:507 (1932).
7. R. L. Myuller, Zh. Fiz. Khim., 6:616 (1935); R. L. Myuller, Acta Physicochim URSS, 2:103 (1935).
8. J. Frenkel', J. Phys. (USSR), 5:31 (1941).
9. R. L. Myuller, The Vitreous State and the Electrochemistry of Glass (Doctoral Dissertation), Leningr. Gos. Univ. (1940).
10. L. Onsager, J. Amer. Chem. Soc., 58:1486 (1936).
11. G. V. Kukolev, Silicon Chemistry and the Physical Chemistry of Silicates, Promstroiizdat, Moscow (1957), p. 597.
12. R. L. Myuller and B. I. Markin, Zh. Fiz. Khim. 5:1272 (1934); R. L. Myuller (Müller) and B. I. Markin, Acta Physicochim. URSS, 7:266 (1934).

V

The Electrical Conductivity of Borosilicates in the Stable State*

1. The author has previously [1] verified in part the theoretical expressions presented earlier for the electrical conductivity of glasses by comparing the calculated with the experimental values. This was based, however, on insufficiently valid theoretical assumptions, and above all the calculations were limited to temperatures lying exclusively below the critical region. The observed conductivity of borosilicates in the stable and labile state may now be explained on the basis of a full theoretical analysis of the expressions for molar conductivity in ionic-covalent systems. We should first compare the experimental and theoretical data for the electrical conductivity of borosilicates in the stable state.

In glasses in the stable state the theoretical molar conductivity is expressed in the form [2]

$$\Lambda_\tau = 10^{P}{}_t \exp\left(\frac{-\Psi_\Phi}{2RT}\right) \frac{\Omega^{-1} \cdot \text{cm}^2}{\text{mole}} ,\qquad(1)$$

where the energy parameter Ψ_Φ is

$$\Psi_\Phi = \Delta\Phi + 2E_A .\qquad(2)$$

$\Delta\Phi$ is here the free energy of the electrolytic dissociation of cations; E_A is the activation energy of the unit translation of a dissociated cation.

With a relatively low metal oxide content in the glass the conductivity is limited by the cations moving in a matrix of nonpolar structural units and in the case of boron glasses we can use the equation

$$P_t = P_a = 3.64 + 0.5 \log \gamma + 2.5 \log z,\qquad(3)$$

where z is the cation valency, γ is the ratio of the non-ionized structural units $[\zeta] = BO_{3/2}$, to the ionized units $[\zeta^-] = B^- O_{4/2}$:

$$\gamma = \frac{[\zeta]}{[\zeta^-]} .\qquad(3a)$$

It is clear that $[\zeta^-] = z[M]$, $[\zeta] = [B] - z[M]$, where $[B]$ is the total concentration of boron atoms (ionized and nonionized) and $[M]$ is the total concentration of cations of a given metal, in mole/ml. These parameters were calculated from the equations

$$[B] = \frac{(\text{wt}\% B_2O_3)\,\delta}{100\,(B_2O_3)/2} = \frac{(\text{wt}\% B_2O_3)\,\delta}{3.48 \cdot 10^3}$$

and

$$[M] = \frac{(\text{wt}\% M_xO)\,\delta}{100\,(M_xO)/x} ,$$

---

*R. L. Myuller, Zh. Tekhn. Fiz., 25(11):1868 (1955).

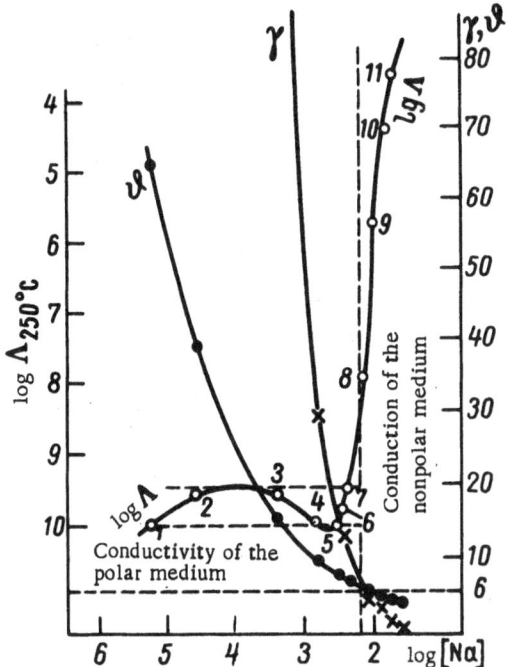

Fig. 1. The curve of the log molar conductivity against log volume concentration of
Na⁺ ions in sodium borate glasses.

where δ is the density of the glass and (B_2O_3)
and (M_xO) are the molecular weights of the
oxides.

Atomic-ionic concentrations in silicate
glasses were determined similarly from densities or molar volumes.

With a sufficiently high metal oxide concentration in the glass, the conductivity is determined by mobile cations and displaced
vacancy "holes" in a medium of polar structural units. Under these conditions we showed
theoretically for borate glasses [2]

$$P_t = P_{\sigma_p} = 3.64 + 2\log z, \qquad (4)$$

and for silicate glasses

$$P_t = P_{\sigma_p} = 3.83 + 2\log z. \qquad (5)$$

As is known, the empirical expression
for the log conductivity,

$$\ln x = -\frac{A}{T} + B, \qquad (6)$$

where A and B are constants, can be expressed
in the molar conductivity form

$$\Lambda_e = 10^{P_e} \exp\left(-\frac{A}{T}\right), \qquad (7)$$

where

$$P_e = 0.4343B - \log[M]. \qquad (8)$$

By comparing Eq. (7) with Eq. (1) we conclude that

$$\Psi_\Phi = 2AR = 3.97\cdot10^{-3}A \text{ kcal/mole}. \qquad (9)$$

This last equality requires experimental justification for the theoretical value of the power parameter P_t. The equality $P_t = P_e$, derived from the comparison of Eqs. (1) and (7), can
only be considered as a rigorously based application of the energy relationship of Eq. (9) if
there is experimental confirmation.

2. The small difference in the parameter P_t in borates and silicates is seen in Eqs. (4)
and (5). In borate glasses, for example, there is also a difference between the conductivity of
the medium formed from nonpolar structural units and that from polar units.

The polar groups are associated and form microinclusions isolated from each other. Conductivity is then limited by the small number of mobile dissociated cations in the nonpolar medium. At some critical concentration of the polar groups they interact and join together. As a
result continuous conductivity provided by the cations in the microdispersed entanglements of
associated polar structural elements becomes easier [3].

In order to characterize the critical concentration of polar groups at which the conductivity in the glass in a nonpolar medium changes to conductivity in a polar medium, it is convenient
to consider the curve of the log of the molar electrical conductivity of the sodium borate glasses,

log Λ (Fig. 1), as a function of the log of the concentration of $[\zeta^-]$ polar groups which are equal to the concentration of metal cations [M], in this particular case, sodium ions [4]. It is reasonable to expect that the abrupt rise in conductivity starts at that critical point at which, on average, one nonpolar unit ceases to find enough space between two polar structural units:

$$\left[B^-O_{\frac{4}{2}} M^+ \right]\left[BO_{\frac{3}{2}} \right]\left[B^-O_{\frac{4}{2}} M^+ \right].$$
$$\xleftarrow{\hspace{1cm}} \theta \xrightarrow{\hspace{1cm}}$$

From the radii of the structural units, polar (R = 1.53 Å) and nonpolar (r = 1.36 Å) [5], we can see that the critical point for the onset of interaction between the polar groups must appear at an average distance between the centers of the polar structural elements of $\theta = 2(R+r) = 5.8$ Å $\cong 6$ Å. General geometrical considerations suggest that the continuous nonpolar layer tween the polar structural units disappears at the critical ratio $\gamma \cong 6$ [6].

Figure 1 also shows, respectively, the curves for the parameters γ and $\theta = ([M] \cdot 6 \cdot 10^{23})^{-1/3}$. 10^8 Å. It follows from Fig. 1 that when the value of $\theta > 6$ Å and $\gamma > 6$ the value of log Λ changes within the narrow limits of 9.5 to 10.0. The molar electrical conductivity undergoes an abrupt increase by almost six orders of magnitude, which is in complete agreement with what has already been indicated in arriving at the critical values of θ and γ.

It will later be conditionally accepted that when $\gamma > 6$ the conductivity is limited by the nonpolar component of the medium and when $\gamma \leq 6$ the conductivity is in fact possible throughout the growing polar component of the medium. Further calculations of P_t have been undertaken in this connection.

3. During the last two decades several workers have obtained fairly extensive experimental results for the steady-state conductivity of borosilicates which satisfy the fundamental requirements of reproducibility and are thus parameters uniquely characterized by the chemical composition of the substance [7]. The available experimental data necessary for verifying that $P_e = P_t$ in vitreous borosilicates in the stable state (below the critical temperature region) are reproduced in Tables 1 to 8. In addition to the usual indications of the composition of the glasses in question, the tables also give the calculated volume concentrations of the cations [M], and the boron [B] or silicon [Si] atoms in mole/ml. These characterize the packing density of the structural elements and may be used in calculating the γ parameter. Later, the values are given for P_{0_p} from Eq. (4) or (5) (z = 1) when $\gamma \leq 6$. When $\gamma > 6$ the power parameter was determined from Eq. (3) in the case of the borates. Finally the tables show the value of the difference, $\Delta = P_e - P_t$.

Tables 1 and 2 cover vitreous borates of univalent alkali oxides and silver oxide. A comparison of the experimental data and the theoretical values for the P parameter indicates that there is good agreement between experiment and theory. The deviation is within the limits $\Delta \leq 2$ and for the majority of glasses within $\Delta \leq 1$, meeting the requirements of a previous paper [7].

The increase in Δ in moving from cesium to silver demands some consideration.

$$\Delta_{Cs} \leqslant \Delta_{Rb} < \Delta_K \leqslant \Delta_{Na} < \Delta_{Li} < \Delta_{Ag}.$$

It is clear that this is caused by the increase in the frequency of the thermal vibrations with a decrease in the radius and atomic weight of the ion, whence there is a simultaneous increase in the polarizability and hence in the bond energy. The latter must reach its greatest value for silver and lithium ions.

TABLE 1. Borate Glasses [4, 8, and 9]

Cation	Glass No.	Chemical characteristics					Electrochemical data				
		Wt. %		100 [M], mole/ml	100 [B], mole/ml	γ	B	P_e	P_t		$\Delta = P_e - P_t$
		M_2O	B_2O_3						P_σ	$P_{\sigma p}$	
Lithium	107	1.59	98.41	0.196	5.23	25.7	6.1	5.3	4.3	—	1.0
	108	2.98	97.02	0.386	5.40	13.0	4.1	4.2	4.2	—	0.0
	109	3.96	96.04	0.522	5.43	9.41	2.8	3.5	4.1	—	—0.6
	111	6.26	93.74	0.844	5.43	5.45	4.6	4.1	—	—	0.5
	17	6.8	93.2	0.925	5.47	4.92	6.5	4.8	—	—	1.2
	112	7.33	92.67	1.00	5.46	4.46	5.1	4.2	—	—	0.6
	113	8.75	91.26	1.22	5.52	3.53	3.8	3.6	—	—	0.0
	114	9.85	90.15	1.40	5.54	2.96	4.0	3.6	—	3.6	0.0
	18	10.2	89.8	1.47	5.57	2.79	6.3	4.5	—	—	0.9
	115	10.8	89.2	1.55	5.52	2.56	4.6	3.8	—	—	0.2
	116	10.58	88.42	1.67	5.52	2.30	4.9	3.9	—	—	0.3
	30	14.40	85.60	2.15	5.52	1.57	5.3	4.0	—	—	0.7
									Δ mean . . . +0.5 ± 0.4		
Sodium	7	7.3	92.7	0.45	5.08	10.3	5.43	4.7	4.1	—	0.6
	8	12.60	87.40	0.79	4.88	5.18	3.28	3.5	—	—	—0.1
	9	16.9 ± 0.5	83.1	1.10	4.83	3.36	4.33	3.8	—	3.6	0.2
	10	25.0	75.0	1.8	4.79	1.67	3.04	3.1	—	—	—0.5
	11	30.8	69.2	2.3	4.60	1.00	3.45	3.1	—	—	—0.5
									Δ mean . . . —0.1 ± 0.3		
Potassium	52	19.38	80.62	0.844	4.74	4.63	4.14	3.9	—	—	0.2
	53	25.28	74.72	1.13	4.52	3.00	3.40	3.4	—	—	—0.2
	54	28.86	71.14	1.31	4.37	2.34	4.25	3.7	—	—	0.1
	55	33.01	66.99	1.52	4.30	1.83	3.91	3.5	—	3.6	—0.1
	56	37.53	62.47	1.76	4.07	1.31	3.45	3.2	—	—	—0.4
	57	39.7	60.3	1.88	3.86	1.05	3.60	3.3	—	—	— 0.3
	58	41.29	58.71	1.98	3.82	0.93	3.80	3.3	—	—	—0.3
									Δ mean. . . . —0.1 ± 0.3		
Rubidium	59	11.5	88.5	0.246	5.08	19.7	—0.9	2.2	4.3	—	—2.1
	60	31.2	68.8	0.76	4.50	4.9	6.0	4.7	—	3.6	1.1
									Δ mean . . . . ± 1.6		
Cesium	61	10.7	84.3	0.23	4.90	20.6	—0.9	2.2	4.3	—	—2.1
	62	34	66	0.58	4.53	6.8	1.2	2.7	4.1	—	—1.9
									Δ mean . . . . —2.0		

Results relating to sodium conductivity in the polar medium of barium borate are considered in Table 3. The calculation of the γ parameter allowed for the polar structural elements of both the sodium and barium types:

$$\gamma = \frac{[B] - \sum z_i [M_i]}{\sum z_i [M_i]},$$

where z_i is the valency of the M_i cation. The deviation Δ in these glasses is also confined within the prescribed limits.

TABLE 2. Borate Glasses [10]. Silver Cation

Glass No.	Wt. %		100 [M], mole/ml	100 [B], mole/ml	γ	B	P_e	P_t		Δ
	M_2O	B_2O_3						P_σ	P_{σ_p}	
145	9.90	90.1	0.176	5.35	29.4	9.26	6.8	4.4	—	2.4
146	15.08	84.92	0.284	5.30	17.7	7.42	5.8	4.3	—	1.5
147	19.87	80.17	0.396	5.33	12.4	6.82	5.4	4.2	—	1.2
148	25.44	74.56	0.544	5.32	8.8	6.55	5.1	4.1	—	1.0
149	31.56	68.44	0.724	5.23	6.25	5.72	4.6	4.0	—	0.6
150	41.74	58.26	1.10	5.11	3.64	5.68	4.4	—	3.6	0.8

Δ mean..... $+1.5 \pm 0.5$

TABLE 3. Borate Glasses [11]. Sodium Cation, 100 [Ba] = 0.82, $P_{\sigma_p} = 3.6, \Delta = 0.8 \pm 0.3$

Glass No.	Wt. %			100 [Na], mole/ml	$100 \Sigma z_i \frac{[M_i]}{}$, mole/ml	100 [B], mole/ml	γ	B	P_e	Δ
	Na_2O	BaO	B_2O_3							
75	5.20	40.33	54.47	0.52	2.16	4.85	1.24	5.61	4.7	1.1
76	10.45	39.8	49.75	1.08	2.72	4.58	0.68	4.71	4.0	0.4
77	12.8	40.5	36.7	1.32	2.96	3.38	0.14	6.10	4.5	0.9

TABLE 4. Borate Glasses [11]. Barium Cation

Glass No.	Wt. %		100 [Ba], mole/ml	100 [B], mole/ml	γ	B	P_e	P_{σ_p}	Δ
	BaO	B_2O_3							
67a	54.1	45.9	1.33	4.88	0.84	2.60	2.9	4.2	—1.3
68	56.5	43.5	1.36	4.62	0.70	4.72	3.9	4.2	—0.3
69	59.8	40.2	1.44	4.28	0.49	5.05	4.0	4.2	—0.2

Δ mean.... -0.6 ± 0.5

4. In Tables 4 and 5 barium and zinc borate glasses are considered. In zinc glasses No. 79 and 80 all the structural units are ionized, some of them two-fold:

In Table 5, therefore, as also in Tables 6 and 7, instead of the γ parameter (inconvenient in the case of doubly ionized structural elements), parameters are introduced which character-

TABLE 5. Borate Glasses [11]. Zinc Cation

Glass No.	Wt. %		100 [Zn], mole/ml	100 [B], mole/ml	α'	α''	B	P_e	P_{σ_p}	Δ
	ZnO	B_2O_3								
78	53.8	46.2	2.18	4.38	1.00	—	0.64	1.94	4.2	−2.3
79	56.3	43.7	2.23	4.02	—	0.11	1.45	2.28	4.2	−2.0
80	64.5	35.5	2.93	3.77	—	0.56	1.70	2.27	4.2	−2.0

Δ mean.... −2.0±0.2

TABLE 6. Silicate Glasses [12,13]. Sodium Cation, $P_{\sigma_p} = 3.83$

Glass No.	Mole %		Molar vol., ml	100 [Na], mole/ml	100 [Si], mole/ml	α'	α''	B	P_e	Δ
	Na_2O	SiO_2								
P128	8.4	91.6	26.5	0.63	3.46	0.18	—	2.78	3.4	−0.4
P127	11.8	88.2	26.2	0.90	3.37	0.27	—	3.00	3.4	−0.4
P126a	15.2	84.8	25.7	1.18	3.30	0.36	—	6.86	3.6	−0.2
P124	19.2	80.8	25.3	1.52	3.19	0.48	—	3.82	3.5	−0.3
P123	22.7	77.3	25.0	1.82	3.10	0.59	—	3.77	3.4	−0.4
P122	25.4	74.7	24.9	2.05	3.00	0.68	—	4.19	3.5	−0.3
P121	26.4	73.6	24.8	2.13	2.98	0.71	—	4.60	3.7	−0.1
P120	27.2	72.8	24.7	2.2	2.95	0.74	—	4.40	3.6	−0.2
P119	29.7	70.3	24.6	2.4	2.85	0.84	—	4.51	3.6	−0.2
P118	32.9	67.1	24.4	2.7	2.76	0.98	—	4.79	3.7	−0.1
P117a	35.7	64.3	24.3	2.9	2.65	—	0.09	4.85	3.7	−0.1
P116a	39.3	60.7	24.1	3.3	2.52	—	0.31	5.98	4.1	0.3
P115a	45.5	54.5	24.0	3.8	2.27	—	0.68	5.38	3.8	0.0
P114	49.8	50.2	23.9	4.2	2.10	—	1.00	5.20	3.6	−0.2

Δ mean.... −0.2±0.1

TABLE 7. Silicate Glasses [14]. Potassium Cation

Glass No.	Mole %		α'	α''	B
	K_2O	SiO_2			
9	17.80	82.20	0.43	—	4.1
8	18.91	81.09	0.47	—	4.1
7	20.70	79.30	0.52	—	3.2
6	23.80	76.20	0.63	—	3.7
5	27.98	72.02	0.77	—	3.7
4	29.68	70.32	0.84	—	2.5
3	32.60	67.40	0.97	—	2.5
2	36.98	63.02	—	0.18	2.8
1	37.96	62.04	—	0.23	2.8

ize the degree of ionization of the covalent glass structure. The degree of single ionization is determined from

$$\alpha' = \frac{[\zeta^-]}{[\zeta^-] + [\zeta]} = \frac{z[M]}{[B]} \quad \text{(borates)} \quad \text{or} \quad \alpha' = \frac{z[M]}{[Si]} \quad \text{(silicates)}. \tag{10}$$

The degree of double ionization (after the single ionization has been completed in all the structural groups) in the borates is determined from

$$\alpha'' = \frac{[\zeta^=]}{[\zeta^=] + [\zeta^-]} = \frac{z[M] - [B]}{[B]} = \frac{z[M]}{[B]} - 1 \tag{11}$$

TABLE 8. Borosilicate Glasses [15]

Cation	Glass No.	Wt. %			Density	100 [M], mole/ml	100 [B], mole/ml	100 [Si], mole/ml	100 [B] + 100 [Si], mole/ml	γ	B	P_e
		M_2O	B_2O_3	SiO_2								
Sodium	300	10.0	62.9	27.1	2.136	0.69	3.86	0.96	4.82	6.00	3.90	3.85
$(P_e = 3.6 \pm 0.2,\ P_{\sigma_p} =$	301	15.0	59.4	25.6	2.244	1.09	3.82	0.96	4.78	3.38	3.94	3.67
$= 3.83,\ \Delta = -0.2 \pm 0.2)$	302	17.5	57.6	24.9	2.290	1.30	3.80	0.95	4.75	2.65	4.01	3.63
	303	20.0	55.9	24.1	2.355	1.52	3.79	0.95	4.74	2.12	3.75	3.45
	304	25.0	52.4	22.6	2.440	1.97	3.67	0.92	4.59	1.33	3.55	3.25
	305	30.0	48.9	21.1	2.470	2.39	3.47	0.86	4.33	0.81	3.96	3.34
	306	35.0	45.4	19.6	2.474	2.79	3.22	0.80	4.02	0.44	5.12	3.77
Potassium	307	20.0	55.9	24.1	2.170	0.92	3.49	0.87	4.36	3.74	3.06	3.37
$(P_e = 3.3 \pm 0.1,\ P_{\sigma_p} =$	308	25.0	52.4	22.6	2.242	1.19	3.38	0.84	4.22	2.54	3.31	3.36
$= 3.83,\ \Delta = -0.5 \pm 0.1)$	309	30.0	48.9	21.1	2.308	1.47	3.24	0.81	4.05	1.76	3.06	3.16
	310	35.0	45.4	19.6	2.372	1.76	3.08	0.77	3.85	1.19	3.32	3.20
	311	40.0	41.9	18.1	2.391	2.03	2.88	0.72	3.60	0.77	4.10	3.47
	312	45.0	38.4	16.6	2.393	2.28	2.64	0.66	3.30	0.45	3.55	3.18

and analogously in silicates. The silicates, although the calculation is identical, differ in structure from the borates. The single ionization of boric oxide structural elements when a metal oxide is introduced is accompanied by an increase of one third in the number of valence bonds between the units. The single ionization of silica structural units, however, leads to the rupture of a quarter of their connecting valence bonds [1]. On the double ionization of silica to give the $SiO_{2/2}O_2^{2-}$ structural unit, two of the connecting valence bonds are broken, but in boric oxide $(B^-O_{3/2}O^-)$ only the first bond is broken. In the case of divalent cations the theoretical value for the power parameter P_{σ_p} in the case of borates is calculated from Eq. (4):

$$P_{\sigma_p} = 3.64 + 2 \log 2 = 4.24.$$

The relatively small values of Δ in Tables 4 and 5 confirm, as before, the theoretical values for the P_t power parameter.

There is extremely good agreement between the experimental and theoretical values for silica (Table 6) and for silica-boric oxide (Table 8) glasses.

Unfortunately, many experimental results for the electrical conductivity cannot be used for a strict comparison with theoretical values. This is true, firstly, in those cases where only relative values of conductivity were determined without eliminating the concentration polarization close to the electrode, as for example [16]. A strict comparison is made difficult when the proportion of the components is given only as a percentage instead of a volume concentration [17]. Specific conductivity, like dielectric loss, refers to a unit of volume and is therefore a parameter proportional to volume concentrations of the mobile cations. Volume molar concentration is often neglected in characterizing substances by conductivity studies. In particular, no volume concentrations were given for the potassium silicate glasses shown in Table 7, but we can show that to an order of magnitude the data in Table 7 confirms the theoretical calculations. In fact, the density of potassium silicate glasses is close to three and for example, for the glasses Nos. 1 and 9 the values of P_e from Eq. (8) are 2.7 and 3.6 respectively with the theoretical value $P_{\sigma_p} = 3.8$ ($\Delta = \pm 0.8$). It is clear that glasses Nos. 1-8 give the same result.

TABLE 9. Borate Glasses [4, 8, 9]

Cation	Glass No.	100 [M], mole/ml	A	Ψ_Φ kcal/mole
Lithium	107	0.196*	18080	71.7
	108	0.386*	16580	65.8
	109	0.522*	14920	59.3
	111	0.844	14230	56.5
	17	0.925	15000	59.5
	112	1.00	13920	55.2
	113	1.22	12000	47.5
	114	1.40	11370	45.0
	18	1.47	11900	47.3
	115	1.55	10800	42.8
	116	1.67	10350	41.0
Sodium	30	2.15	9500	37.7
	7	0.45*	17100	67.9
	8	0.79	13800	55.1
	9	1.10	11600	46.0
	10	1.8	9103	36.2
	11	2.3	8204	32.5
Potassium	52	0.84	15100	60.0
	53	1.13	12050	47.8
	54	1.31	11570	45.8
	55	1.52	10200	40.5
	56	1.76	9400	37.3
	57	1.88	8970	35.6
	58	1.98	8850	35.0
Rubidium	59	0.25*	16600	65.8
	60	0.76	16900	67.0
	61	0.23*	16600	65.8
	62	0.58*	16450	65.2

* $\gamma > 6$

5. Following the strict confirmation of an agreement between the experimental and theoretically calculated P_e and P_t parameters in various vitreous borosilicates at temperatures below the critical region, we can now consider the fundamental theoretical quantitative interpretation presented in [2] of the physical meaning of the pre-exponential term 10^{Pe} in the empirical Eq. (7) for borosilicates in the stable state. The identity thus established between Eqs. (1) and (7) allows us to consider Eq. (9) as valid. In Tables 9–12 values of the Ψ_Φ parameter from Eq. (2) for the case of the glasses in question, obtained from the values of the empirical temperature coefficients A from Eq. (6) are given.

In borate glasses with the highest lithium and sodium content, values of Ψ_Φ are found to be close to 40 kcal/mole (Table 9). The provisional theoretical results [7, Table 1] are in good agreement. The theoretical value exceeds the experimental only in the case of potassium. Allowing for the approximate character of the theoretical calculations we can see here, too, a confirmation of the general concepts, based on the deduction from the molar conductivity equation [2], of electrolytic dissociation in ionic–covalent media. This implies that the activation energy E_A is relatively small which allows us to equate, to a first approximation, the energy Ψ_Φ with the free energy of electrolytic dissociation $\Delta\Phi$ [2].

The numerical values of the Ψ_Φ parameter will be treated in greater detail later.

6. To conclude this investigation of electrical conductivity at temperatures below the critical region we must remember that, in all the studies of vitreous borosilicates, the empirical coefficient A does not depend under these conditions on the temperature. This is indicative of the temperature independence of Ψ_Φ and therefore of $\Delta\Phi$.

TABLE 10. Borate Glasses [10, 11]

Cation	Glass No.	100 [M], mole/ml	A	Ψ_Φ, kcal/mole
Silver	145	0.18*	19550	77.6
	146	0.28*	17950	71.2
	147	0.40*	17000	67.5
	148	0.54*	16100	64.4
	149	0.72*	19980	58.8
	150	1.10	12075	48.3
Sodium in a barium glass medium	75	0.52	19200	76.2
	76	1.08	15000	59.5
	77	1.32	14100	56.0
Barium	67a	1.33	19600	77.8
	68	1.36	20500	80.3
	69	1.44	19900	79.0
Zinc	78	2.18	17630	70.0
	79	2.23	17600	69.8
	80	2.93	17400	67.5

* $\gamma > 6$.

TABLE 11. Silicate Glasses [12, 14]

Cation	Glass No.	100 [M], mole/ml	A	Ψ_Φ, kcal/mole
Sodium	P128	0.63	9280	36.8
	P127a	0.90	8800?	34.9
	P126a	1.18	8820	35.0
	P124a	1.52	8480	33.4
	P123	1.82	7940	31.5
	P122	2.05	7830	31.1
	P121	2.13	8080	32.0
	P120	2.2	7750	30.6
	P119	2.4	7530	29.8
	P118	2.7	7330	29.1
	P117a	2.9	7250	28.7
	P116a	3.3	6930	27.5
	P115a	3.8	6850	27.2
	P114	4.2	6470	25.6
Potassium	9	—	10250	40.7
	8	—	9850	39.1
	7	—	9170	36.4
	6	—	8600	34.1
	5	—	8350	33.1
	4	—	7840	31.1
	3	—	7350	29.2
	2	—	7080	28.1
	1	—	6680	26.5

It follows that, in glasses in the stable state, electrolytic dissociation is not accompanied by a change in the entropy of the system; $\Delta S = 0$.* This confirms the freezing of the thermodynamically nonequilibrium ionic-covalent systems in borosilicates below the critical temperature region and the absence of structural changes in them during electrolytic dissociation.

The comparison of experiment with theory confirms the fundamental value of applying thermodynamic laws to phenomena connected with cation distribution in stabilized glasses. This

*We must remember that all these statements are valid only when $P_e \approx P_t$. If $P_e > P_t$ then this indicates, according to Myuller, that there is an increase in the entropy of the dissociation process and therefore a rise in Ψ_Φ and $\Delta\Phi$ with an increase in temperature (Editor's Note).

TABLE 12. Borosilicate Glasses [15]

Cation	Glass No.	100 [M], mole/ml	A	Ψ_Φ; kcal/ mole
Sodium	300	0.69	13700	54.2
	301	1.09	11720	46.6
	302	1.30	10960	43.5
	303	1.52	10120	40.2
	304	1.97	9020	35.8
	305	2.39	8410	33.4
	306	2.79	8220	32.6
Potassium	307	0.92	13930	55.3
	308	1.19	12260	48.7
	309	1.47	11040	43.9
	310	1.76	10390	41.3
	311	2.03	9530	37.9
	312	2.28	8470	33.6

is an indication that, in an essentially thermodynamically nonequilibrium frozen system of ionic-covalent structural elements, cation processes occur freely, ensuring equilibrium in the electrolytic dissociation of the cations.

Conclusions

1. By using the experimental data, obtained under strictly valid conditions, relating to the electrical conductivity of vitreous borosilicates below the critical temperature region, we have demonstrated the validity of the calculated values of the pre-exponential term in the theoretical equation for the molar conductivity of borosilicates. For stable borosilicates the temperature coefficient A in the empirical conductivity equation was shown to be an energy term which could be used to calculate the parameter $\Psi_\Phi = 2AR$.

2. Considerations are presented in favor of the view that a change from conductivity in the nonpolar component of borosilicates to conductivity in the polar structure occurs as the ratio of the nonpolar structural units to the polar approaches roughly, six.

3. The experimental confirmation of the theoretical conductivity equation indicates the validity of using this as a basis for the assumption that the electrolytic dissociation of polar structural elements is in thermodynamic equilibrium. At the same time the temperature independence of the energy Ψ_Φ, and of the free energy of the electrolytic dissociation, demonstrates the absence of an entropy change in the system on dissociation ($\Delta S = O$). This confirms the freezing of the thermodynamically nonequilibrium ion-covalent system of borosilicates below the critical range of temperatures and the absence of local structural changes in them on electrolytic dissociation.

References

1. R. L. Myuller, Zh. Fiz. Khim., 6:616 (1935); Izv. Akad. Nauk SSSR, ser. fiz., 4:607 (1940); •this volume, p. 170; The Vitreous State and the Electrochemistry of Glass (Doctoral Dissertation), Leningr. Gos. Univ. (1940); R. L. Myuller, Acta Physicochim. URSS, 2:103 (1935).

2. R. L. Myuller, Zh. Tekhn. Fiz., 25:246 (1955), • this volume, p. 24.

3. R. L. Myuller (Müller), Nature, 129:507 (1932).

4. S. A. Shchukarev and R. L. Myuller, Zh. Fiz. Khim., 1:625 (1930); S. A. Shchukarev and R. L. Myuller (Müller), Z. Phys. Chem., A150:439 (1930).

5. R. L. Myuller, Zh. Tekhn. Fiz., 25:1556 (1955), • this volume, p. 33.

6. R. L. Myuller (Müller), Phys. Z. Sowjetunion, 1:407 (1932).

7. R. L. Myuller, Zh. Tekhn. Fiz., 25:1567 (1955), ● this volume, p. 43.

8. R. L. Myuller and B. I. Markin, Zh. Fiz. Khim., 5:1272 (1934); R. L. Myuller and B. I. Markin, Acta Physicochim. URSS, 1:266 (1934).

9. L. R. Takking and N. P. Shchegoleva, Uch. Zap. Leningr. Gos. Univ., No. 108, p. 17, 1949.

10. B. I. Markin, Byull. Vses. Khim. Obshchestva im. Mendeleeva, No. 5 (1940); Zh. Khim. Obshchestva, 11:285 (1941).

11. R. L. Myuller and B. I. Markin, Zh. Fiz. Khim., 7:592 (1936); R. L. Myuller and B. I. Markin, Acta. Physicochim. URSS, 4:471 (1936).

12. E. Seddon, E. Tippett, and W. E. S. Turner, J. Soc. Glass Technol., 16:459 (1932).

13. C. J. Peddle, J. Soc. Glass Technol., 4:9 (1920); F. Wilkinson and W. E. S. Turner, J. Soc. Glass Technol., 15:185 (1931); W. Blitz and F. Weibke, Z. anorg. allg. Chem., 203:345 (1932).

14. A. Ya. Kuznetsov and I. G. Mel'nikova, Zh. Fiz. Khim., 24:1204 (1950).

15. B. I. Markin, Zh. Tekhn. Fiz., 22:932 (1952).

16. A. F. Val'ter, M. A. Gladkikh, and K. I. Martyushov, Zh. Tekhn. Fiz., 10:1593 (1940).

17. G. I. Skanavi, Dielectric Physics, GITTL, Moscow (1949).

VI

The Electrical Conductivity of
Borosilicates in the Labile State*

1. When borosilicates are heated above T_0, the lower limit of the critical temperature region, an abrupt increase is observed in the coefficients A and B in the empirical conductivity equation

$$\ln \varkappa = -\frac{A}{T} + B. \tag{1}$$

The value of the power parameter, in the well known expression for molar conductivity

$$\Lambda = 10^{P_e} \exp\left(-\frac{A}{T}\right) \tag{2}$$

$$P_e = 0.4343B - \log[M] \tag{3}$$

begins to exceed the theoretically calculated value by a considerable amount

$$P_{a_p} = a + 2.5\log z + 0.5\log \gamma, \tag{4}$$

where a is 3.64 in borate and 3.83 in silicate glasses; z is the electrovalence of the conducting ion; and

$$\gamma = \frac{[\zeta]}{[\zeta^-]} \tag{5}$$

satisfies the ratio of the concentration of nonpolar structural units $[\zeta]$ to that of the polar units $[\zeta^-]$ [1]. In borates, $[\zeta]$ = [B] - z[M] and in silicates $[\zeta]$ = [Si] - z[M]; $[\zeta^-]$ = z[M]. Here [B], [Si], and [M] are the volume concentrations of boron atoms, silicon atoms, and conducting cations, respectively, in moles per ml.

Table 1 shows the experimental data for borate glasses in the critical temperature region.

We see that the experimentally determined pre-exponential term is 10^4 times, and in individual cases (Glasses Nos. 44-47, 49, and 146) even 10^{10}-10^{17} times, greater than the theoretically calculated values.

2. On the transition into the critical temperature region there is a considerable increase in the value of A (Eq. (1)) together with a rise in the value of B. We therefore assume that a breakdown of the structure occurs here and is accompanied by a change in the activation energy of the ions [3; 4, p. 295].

According to studies of the thermal capacity, electrolytic dissociation in stable state glasses must occur without any local changes in the degree of order of the structure, i.e. there

---

*R. L. Myuller, Zh. Tekhn. Fiz., 25:2428 (1955).

TABLE 1. Borate Glasses [2]

Cation	Glass No.	Wt.%		100 [M], mole/ml	100 [B], mole/ml	γ	B	P_e	P_σ	$P_e - P_\sigma$
		M_2O	B_2O_3							
Lithium	13	0.1	99.9	$1,2 \cdot 10^{-2}$	5.22	434	17.1	11.3	5.0	6.3
	14—15	0.55	99.45	$6,6 \cdot 10^{-2}$	5.07	76	22.3	12.9	4.6	8.3
	107	1.59	98.41	0.196	5.23	26	18	10.5	4.3	6.2
	16	2.31	97.69	0.294	5.35	17	11.5	7.5	4.3	3.2
	108	2.98	97.02	0.386	5.40	13	13	8	4.2	3.8
Sodium	1	<0.1	>99.99	$<5.8 \cdot 10^{-4}$	5.17	>9000	12.8	10.8	5.6	5.2
	2	0.047	99.95	$2.7 \cdot 10^{-3}$	5.09	1880	14.6	10.9	5.3	5.6
	3	0.71	99.29	$4.3 \cdot 10^{-2}$	5.32	123	21.4	12.6	4.7	7.9
	4	2.78	97.28	0.16	5.00	30	26.3	14.2	4.4	9.8
	5	5.27	94.73	0.31	4.97	15	14.1	8.6	4.2	4.4
	6	5.8	94.2	0.35	5.07	13.5	12.0	7.6	4.2	3.4
	7	7.3	92.7	0.45	5.08	10.3	5.4	4.7	4.1	0.6
Potassium	44	1.24	98.76	$4.8 \cdot 10^{-2}$	5.17	107	} 30	16.3	4.7	11.6
	45	3.36	96.64	} $1.32 \cdot 10^{-1}$	5.08	37.5		15.8	4.4	11.4
	45a	3.34	96.66						4.3	11.5
	46	5.03	94.97	0.20	5.14	24.7	} 13.9	16.6	4.3	12.3
	47	6.50	93.50	0.262	5.13	18.6		16.5	4.3	12.2
	48	9.18	90.82	} 0,37	5.00	12.5	} 10.0	12.4	4.2	8.2
	48a	8.95	91.05							
	49	11.03	88.97	0.46	5.00	9.9		12.3	4.1	8.2
	50	15.20	84.80	0.65	4.88	6.5	} 7	5.2	4.0	1.2
	51	15.72	84.28	0.67	4.83	6.3		5.2	4.0	1.2
Rubidium	59	11.5	88.5	0.25	5.08	19.7	41.6	20.6	4.3	16.3
	60	31.2	68.8	0.76	4.50	4.9	6	4.7	4.0	0.7
Cesium	61	15.7	84.3	0.23	4.98	20.6	20.2	11.4	4.3	7.1
	62	34	66	0.58	4.53	6.8	12.8	7.7	4.1	3.6
Silver	145	9.90	90.1	0.18	5.17	29.4	27.95	14.8	7.3	7.5
	146	15.08	84.92	0.28	5.02	17.7	31.3	16.1	6.3	9.8
	147	19.87	80.17	0.40	4.93	12.4	24.8	13.2	5.9	7.3
Na + Ba	74	3.28	56.45	0.33	5.03	1.56	6.55	9.0	3.7	5.3

can be no change in the entropy of the system [1]. It is different for glasses in the labile state. Here, local structural changes occur on dissociation due to the disappearance, in the critical region, of the rigidity of the valence bonds. The degree of structural order changes locally on dissociation as a result of the development of the $(B^-O_{4/2})$ or $(SiO_{3/2}O^-)$ structural units and dissociated positive cations with central charges from the $(B^-O_{4/2}M^+)$ or $(SiO_{3/2}O^-M^+)$ structural units. This causes both a change in the free energy of dissociation $\Delta\Phi$ (and consequently in the energy parameter Ψ_ϕ) and a change in the entropy of the medium on dissociation: $\Delta S > O$. We can thus state definitely that in the expression for the molar conductivity of glasses in the stable state [1]

$$\Lambda = 10^{P_\sigma} \exp\left(-\frac{\Psi_\Phi(T)}{2RT}\right) \qquad (6)$$

the pre-exponential term is independent of temperature and in the labile state retains the value observed in the stable state. Only the energy parameter $\Psi_\Phi(T)$ is variable and changes with temperature in the critical region as a result of a change in the free energy of electrolytic dissociation $\Delta\Phi$.

3. From the experimental data for the specific conductivity $\varkappa$ at various temperatures within the critical region and using the theoretical values of the power term P_σ it was possible to determine the values of the $\Psi_\Phi(T)$ parameter. From Eq. (6) it follows that

$$\Psi_\Phi(T) = 2R(2.3 P_\sigma - \ln\Lambda)T = 3.97 (B_\sigma - \ln\varkappa)T \text{ kcal/mole,} \qquad (7)$$

$$\Psi_\Phi(T) = 9.15 (0.4343 B_\sigma - \log\varkappa)T \text{ kcal/mole,} \qquad (8)$$

where $B_\sigma = P_\sigma + \log[M]$, while $\Lambda = \varkappa/[M]$ [1].

TABLE 2. Borosilicate Glasses [2]

Glass No.	B_σ	T, °K	log ϰ	$\Psi_\Phi(T)$, kcal/mole	$\Delta S(T)$, kcal/mole	$\Psi_H(T)$, kcal/mole
107	2.28	480	13.7	64.5	—	—
		492	—	64.3	16.7	72.5
		504	62.9	64.1	—	—
		540	—	62.7	38.4	83.4
		577	10.6	61.3	—	—
108	2.74	479	13.2	63.0	—	—
		499	—	62.8	7.3	66.4
		520	12.0	62.7	—	—
		539	—	62.4	13.1	69.5
		558	11.0	62.2	—	—
		584	—	60.1	40.4	83.7
		610	9.4	59.1	—	—

We shall call this strict method of determining $\Psi_\Phi(T)$ "method I."

Unfortunately it has limited application, since there is only a small number of published papers which quote the complete numerical experimental results for the electrical conductivity. The normal practice is to limit such data to graphical reproductions of the experimental curves for the temperature dependence of electrical conductivity. This of course makes it impossible to recover, with the necessary accuracy, the numerical values for the specific conductivity at various temperatures.

Lithium and sodium borate glasses are exceptions as the individual numerical values of their conductivity have been given by the authors in [2]. Tables 2 and 3 show a few of these results for the critical temperature regions, together with the values of the $\Psi_\Phi(T)$ parameter calculated from Eqs. (7) and (8) and the values derived from them for $\Delta S(T)$ and $\Psi_H(T)$ following [1]:

$$\Delta S(T) = -\frac{\partial \Psi_\Phi(T)}{\partial T} = -\frac{\Delta \Psi_\Phi(T)}{\Delta T} \quad (\Delta T \text{ small}) \tag{9}$$

$$\Psi_H(T) = \Psi_\Phi(T) + T\Delta S(T). \tag{10}$$

All three parameters are functions of temperature. In fact the data in Tables 2 and 3 conform this.

A more careful examination of Tables 2 and 3 indicates that the $\Psi_H(T)$ parameter determined from the free energy of dissociation, in all the glasses without exception, decreases with an increase in temperature. As for the ΔS and $\Psi_H(T)$ parameters, with a concentration of $[M] < \cdot 10^{-5}$ (glasses Nos. 1 and 2) these decrease with temperature; with a concentration of $[M] = 4 \cdot 10^{-4}$ (glass No. 3) there is no change, and with a concentration of $[M] > 1.5 \cdot 10^{-3}$ (glasses Nos. 4-7, 107, and 108) they increase. This principle can be explained as follows. All the data in question relate to the same comparatively narrow temperature interval, about 500-600°K. As is known, when the alkali oxide content in borate glasses is increased (increase in [M]) the critical region is shifted towards higher temperatures. In borate glasses this region is comparatively narrow and lies within some 100° [7]. This region is that in which the entropy effect $\Delta S(T)$ increases at the beginning of the critical region and drops to zero at the end (the region of melting).

The term $\Delta S(T)$ appears, and increases at first because of the disappearance of the rigid valence bonds and the consequent possibility of local changes in the degree of order when dissociation occurs. At the end of the critical region, near the melting temperature, $\Delta S(T)$ must fall to zero because no change occurs in the degree of local order when dissociation occurs.

TABLE 3. Sodium Borate Glasses [2]

Glass No.	B_σ	T, °K	$-\ln \varkappa$	$\Psi_\Phi(T)$	$\Delta S(T)$	$\Psi_H(T)$	Glass No.	B_σ	T, °K	$-\ln \varkappa$	$\Psi_\Phi(T)$	$\Delta S(T)$	$\Psi_H(T)$
1	−0.55	497	37.62	73.0	—	—	4	2.3	513	30.5	66.8	—	—
		514	—	72.0	58	101.8			529	—	65.5	78.2	106.8
		531	34.29	71.0	—	—			545	27.4	64.3	—	—
		543	—	70.6	31	87.4			557	—	63.0	104.0	121.0
		554	32.40	70.3	—	—			570	25.0	61.7	—	—
		566	—	69.9	32	88.0	5	2.50	513	29.7	65.7	—	—
		579	30.80	69.5	—	—			523	—	65.3	35.0	83.6
2	0.30	499	34.99	70.0	—	—			533	28.2	65.0	—	—
		521	—	68.7	57	98.3			547	—	64.3	48.3	90.7
		543	31.00	67.5	—	—			562	26.0	63.6	—	—
		555	—	66.8	62.5	102.2	6	2.62	526	27.8	63.5	—	66.1
		567	28.94	66.0	—	—			546	—	63.4	4.9	—
		572	—	65.8	33.3	84.9			567	25.5	63.3	—	110.1
		577	28.44	65.7	—	—			579	—	62.3	82.6	—
3	1.68	511	31.0	66.3	—	—			590	23.6	61.4	—	—
		538	—	64.3	74.2	104.2	7	2.64	463	31.5	62.8	—	66.9
		565	26.1	62.3	—	—			482	—	62.5	13.3	—
		581	—	61.1	74.2	104.1			500	29.0	62.3	—	68.9
		596	23.6	60.0	—	—			530	—	61.9	13.3	—
									559	25.1	61.5	—	—

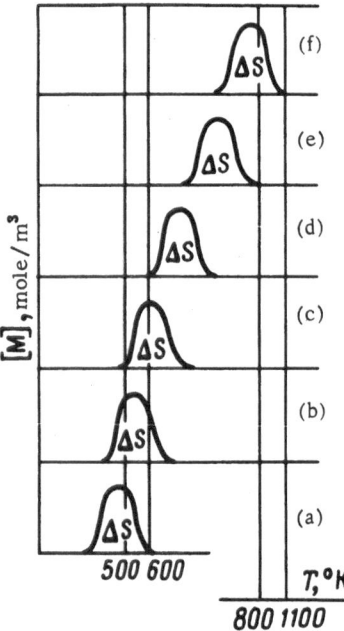

Fig. 1. The shift of the temperature region of the entropy effect ΔS during electrolytic dissociation in borate glasses with a change in cation concentration: a) glass Nos. 1, 2, $[M] < 3 \cdot 10^{-5}$; b) glass No. 3, $[M] = 4 \cdot 10^{-4}$; c) glass Nos. 4-7, 107, 108, $4 \cdot 10^{-4} < [M] < 7 \cdot 10^{-3}$; d) glass Nos. 8-9; e) glass No. 10; f) glass No. 11, $[M] = 2.3 \cdot 10^{-2}$.

On melting the geometric ordering of mutually oriented structural elements disappears and the disordered thermal vibrations become sufficiently intense to eliminate any effect of dissociation on the geometrical order of the system.

This is illustrated in Fig. 1. In the narrow region of the temperatures studied, 500-600°K, in glasses Nos. 1 and 2 (in fact, boric oxide) only the falling branch of the entropy effect was detected from their electrical conductivity. The melting temperature (567°K) confirms this interpretation of the phenomenon. In glass No. 3 there is no noticeable change in ΔS since the higher central section of the entropy curve, decreasing slightly with temperature, falls in the temperature region under investigation. Glasses Nos. 4-7, 107, and 108, with a higher critical temperature region show only the initial section of the entropy curve in this temperature range. Finally, in all the remaining glasses of a higher alkali oxide content, the entropy effect is presented at temperatures above the region examined.

What has been said of the entropy ΔS applies also to $\Psi_H(T)$. At the end of the critical region the latter must fall, approaching the value of $\Psi_\Phi(T)$. Electrolytic dissociation then becomes "anentropic." An analogous disappearance of the entropy effect above T_m is noted in the viscosity of silicate glasses.

In sodium borate glasses the above account is confirmed by the existence of the falling section of the entropy curve in borax in the vicinity of the melting point (1014°K). Table 4 shows the corresponding calculations from Oka's data [5] and Fig. 1 shows diagramatically the corresponding section of the curve in this glass (No.11) in the high temperature region 800-1100°K.

TABLE 4. Borax [5]

Glass No.	B_σ	T, °K	$-\ln \varkappa$	$\Psi_\Phi(T)$	$\Delta S(T)$	$\Psi_H(T)$
11	3.45	773	6.34	30.0	—	—
		823	—	27.9	41	57.5
		873	4.04	25.9	—	—
		923	—	24.7	24	46.8
		973	2.64	23.5	—	—
		1023	—	23.0	9	32.2
		1073	1.86	22.6	—	—

TABLE 5. Borate Glasses [2]

Cation	Glass No.	T_0,° K	A	Ψ_H	ΔS	$\Psi_\Phi(T)$, kcal/mole	$\Psi_\Phi(T)$, kcal/mole 550 °K	580 °K	600 °K
Lithium	13	520	25000	99.2	58	69	67	65	64
	14—15	520	27200	108.0	79	67	65	63	61
	107	550	24450	97.1	58	65	—	63	62
	16	560	20000	79.5	29	63	—	62	62
	108	570	21600	85.7	35	66	—	66	65
Sodium	1	520	26000	103.0	48	78	76	75	74
	2	520	24710	98.0	51	72	70	69	67
	3	520	26790	106.5	72	69	67	65	63
	4	530	29170	116.0	90	68	66	64	62
	5	560	22450	84.2	40	62	—	61	60
	6	570	22000	87.4	31	70	—	70	69
	7	590	17100	67.9	6	64	—	—	64
Potassium	44—45	520	32000	127.0	105	72	69	66	64
	46—47	530	33000	131.0	112	72	69	66	64
	48—49	550	28000	111.0	75	70	—	68	66
	50—51	580	18500	73.5	11	67	—	—	67
Rubidium	59	550	40000	158.0	149	76	—	72	68
	60	~620	16900	67.0	6	63	—	—	—
Cesium	61	550	28000	111.0	65	75	—	73	72
	62	620	22400	88.9	33	69	—	—	—
Silver	145	560	29900	118.5	69	80	—	78	77
	146	560	33300	131.0	90	81	—	79	77
	147	560	27140	107.5	67	70	—	68	67
Na + Ba	74	590	26400	105.0	49	77	—	—	76

4. As has been noted already, the strict "method I" for determining $\Psi_\Phi(T)$ has at present a limited application because of the incompleteness of the published experimental material on the conductivity of glasses. Therefore we have to resort to the approximation of method II, which involves a calculation based on the empirical values of the A and B coefficients of Eq. (1), which are usually given approximately by authors for glasses in the critical temperature region. While in glasses in the stable state the A and B coefficients do not depend on temperature, in the labile state these coefficients are not constant due to the absence of a strict linear dependence of log conductivity on the reciprocal of the temperature [1]. It is not difficult to prove this if we examine more closely the conductivity curves for Li and K glasses low in alkali [2, 6].

The above analysis of the numerical data for alkali borates confirms this directly.

By applying Eq. (1) to glasses in the labile state the approximate method II makes it possible to carry out a simple calculation using the formulae in [1].

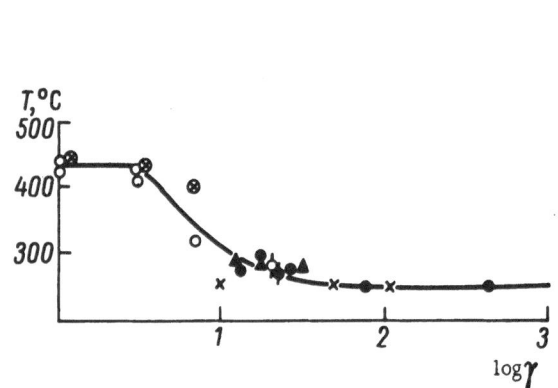

Fig. 2. The curve of the lower limits of the critical temperature region for borate glasses with univalent cations, from data for electrical conductivity [2], viscosity, the coefficient of expansion and elastic vibration.

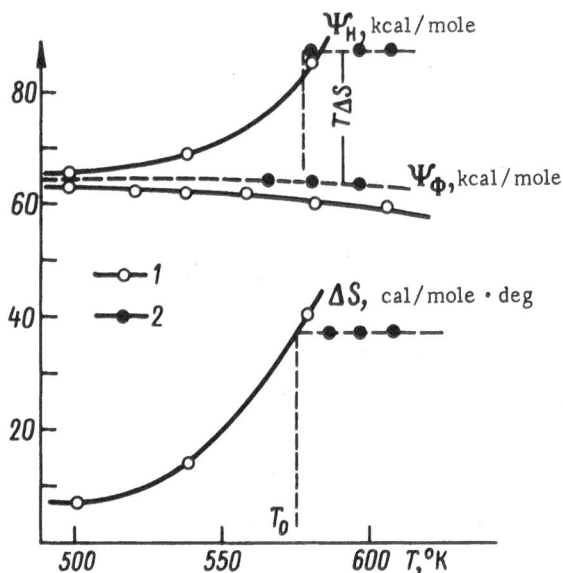

Fig. 3. The dependence of Ψ_Φ, Ψ_H, and ΔS on temperature: 1) calculated by Method I; 2) by Method II.

$$\Delta S = 4.57\,R\,(P_e - P_t) = 9.15\,(P_e - P_\sigma)\,\text{cal/mole} \cdot \text{deg},$$
$$\Psi_H = 2AR = 3.97\,A\,\text{kcal/mole},$$
$$\Psi_\Phi = \Psi_H - T\Delta S\,\text{kcal/mole}. \tag{11}$$

From the data $(P_e - P_\sigma)$ in Table 1 and the approximated empirical values of the A coefficient [2] in Table 5 the parameters are then obtained from Eq. (11). Because there is no precise date for the glasses in question for the lower limits of the critical temperature region T_0, averaged values were taken derived from studies on electrical conductivity, viscosity, coefficient of thermal expansion and elastic vibrations in borate glasses of univalent metals [2, 7] (Fig. 2).

By comparing the results of the calculations in Tables 2 and 3 with those in Table 5 we may note the invariable decrease with temperature of Ψ_Φ for all the glasses in the labile state. Calculations of this parameter using either method I or II give very similar values. Moreover, the values of Ψ_Φ in Tables 2 and 5 near the lower limit of the critical temperature region are close to the values of this parameter in the corresponding glasses in the stable state [1].

There is also a significant deviation in the values of ΔS and Ψ_H in Table 5 from those in Tables 2 and 3. The approximate assumption of the constancy of the A and B coefficients predetermines the constancy of the ΔS and Ψ_H values in Table 5, in accordance with Eq. (11). ΔS and Ψ_H calculated using method II change slowly with temperature and then show a sharp increase at a temperature T_0. This is clearly demonstrated in Fig. 3 by the lithium borate glass No. 108. In this case the values of ΔS and Ψ_H are averaged to give values close to the real values (determined by Method I).

5. The approximate calculation of $\Psi_\Phi(T)$ by a method suggested by Frenkel' [8] will be termed Method III. An additional term is introduced in the expression for the log of the electrical conductivity

$$\ln x = -\frac{A_\sigma}{T} + B_\sigma + \alpha (T - T_0), \tag{12}$$

where α is a constant, and T_0 is the temperature above which the apparent increase is observed in the B coefficient of Eq. (1). In Frenkel's view the coefficient remains unchanged and equal to B, and only A decreases according to $A = A_\sigma - \alpha T (T - T_0)$. The latter expression, as can be easily seen, suggests a quadratic decrease in the parameter $\Psi_\Phi (T) = 2AR = 2R(A_\sigma + \alpha T_0 T - \alpha T^2)$ above some temperature T_0.

The first practical application of Frenkel's expression Eq. (12) to silicate glasses required the introduction of yet a second additional term $\beta (T - T_0')$ [8] into Eq. (12) in a fairly narrow temperature interval (less than 100°). Thus the expression for $\Psi_\Phi (T)$ became more complicated and took the form

$$\Psi_\Phi = 2R \left[A_\sigma + (\alpha T_0 + \beta T_0') T - (\alpha + \beta) T^2 \right], \tag{13}$$

where β is a constant and T_0' is the second reference temperature for the increased growth in conductivity.

The values of the α and β coefficients were calculated graphically and T_0 and T_0' were determined from the dependence of the log of the conductivity of sodium silicate glasses in the labile state on temperature from the data presented by Seddon, Tippett and Turner [8,9]. Table 6 reproduces the values of α, β, T_0, and T_0' for the glasses in question, as well as the A_σ parameters (the values of β_σ have been considered earlier; they confirm the theoretical calculations for glasses in the stable state [1]). Table 6 also shows the $\Psi_\Phi (T)$ values recalculated from Eq. (13) and the values of ΔS and $\Psi_H (T)$ from

$$\Delta S (T) = 2R \left[2(\alpha + \beta) T - (\alpha T_0 + \beta T_0') \right] \tag{14}$$

and

$$\Psi_H (T) = 2R \left[A_\sigma + (\alpha + \beta) T^2 \right], \tag{15}$$

which are derived directly from Eq. (13) using Eqs. (9) and (10).

In considering the values of the parameters in Table 6 we should note that, for all the glasses, there is a decrease in Ψ_Φ and an increase in ΔS and Ψ_H with temperature. This is in accordance with the fact that the data relates to the first section of the critical temperature region some distance from the melting temperature of the compositions in question (only the initial section of the ΔS curves is included).

The calculation by Method III is more accurate according to Frenkel' than Method II, since it involves the temperature dependence of $\Delta S(T)$ and $\Psi_H (T)$. Consequently, as one would expect, its application over a wider temperature range requires the introduction of new supplementary terms and results in awkward calculations from

$$\ln x = -\frac{A_\sigma}{T} + B + \sum_i \alpha_i (T - T_{0,i}). \tag{12a}$$

Above all, this calculation like the calculation by Method I, requires the direct operation on the primary numerical experimental results for the conductivity at various temperatures. From this point of view there is no doubt that Method I described previously is to be preferred to Method III for both accuracy and simplicity.

6. The theory of the electrical conductivity of glass in the labile state allows us to explain the anomalous cross-over of the conductivity curves for borates containing different amounts of the same alkali oxide [2]. Figure 4 shows diagramatically this type of cross-over. Potassium borate glasses No. 44-47 (curve 2) with lower potassium ion concentrations (1000 [M] = 0.5 to 2.6 mole/ml), must have a lower conductivity than glass No. 51 (curve 1) with higher

TABLE 6. Sodium Silicate Glasses [1, 8, and 9]

Glass No.	A_σ	$\Psi_{\Phi o}$, kcal/mole	T_0, °K	1000α	T'_0, °K	1000β	T, °K	ΔS, cal/mole·deg	Ψ_H, kcal/mole	Ψ_Φ, kcal/mole
P128	9280	36.8	667	5.1	—	—	700	14.7	46.9	36.4
							800	18.7	49.6	34.8
P127a	8800	34.9	662	2.3	733	7.1	750	29.4	56.0	33.7
							800	33.0	58.7	32.2
P126a	8820	35.0	662	3.2	733	7.4	750	33.4	58.7	33 8
							800	37.7	62.0	32.0
P124a	8480	33.4	663	1.7	743	9.4	750	34.2	58.3	32.8
							800	38.8	62.0	30.8
P123	7940	31.5	662	1.5	733	3.9	750	17.1	43.7	31.0
							800	19.1	45.2	30.0
P122	7830	31.1	662	1.4	733	8.0	750	29.0	52.0	30.3
							800	32.6	54.8	28.8
P121	8080	32.0	662	1.1	723	6.2	750	22.6	48.5	31.3
							800	25.8	50.4	30.2
P120	7750	30.6	664	2.1	733	4.6	750	20.6	45.7	30.0
							800	23.8	47.6	28.8
P119	7530	29.8	648	1.7	733	6.0	750	23.8	46.8	29.2
							800	27.0	49.2	27.8
P118	7330	29.1	628	1.3	733	8.0	750	28.6	49.8	28.3
							800	32.6	52.6	26.8
P117a	7250	28.7	618	3.2	703	7.8	750	35.8	53.8	25.9
							800	40.1	56.6	24.6
P116a	6930	27.5	669	8.1	693	6.2	700	40.9	55.3	26.7
							750	46.5	59.1	24.6
							800	51.3	63.5	22.2
P115a	6850	27.2	632	4.6	—	—	650	12.3	33.4	27.0
							800	17.7	38.9	24.7
P114	6470	25.6	628	2.3	—	—	650	6.7	29.9	25.6
							800	9.6	32.0	24.3

potassium ion concentrations (1000 [M] = 6.6 mole/ml). This, however, is observed only at temperatures below about 590°K at which the curves intersect. At higher temperatures, glasses Nos. 44–47, relatively low in potassium ions, become much better conductors than glass No. 51 with a high potassium content. It is not difficult to see in Fig. 4 that the intersection is due to a shift of the lower limit of the critical region towards higher temperatures as the alkali oxide content of the borate glass is increased (see Table 5).

The cross–over of the conductivity curves 3 and 4 (Fig. 4) for the sodium borate glasses No. 4 (1000 [M] = 1.6) and No. 5 (1000 [M] = 3.1) respectively, is the results of a more significant entropy effect and a more rapid fall of Ψ_Φ with temperature in the labile state in glass No. 4 than in glass No. 5 (see Table 3). In spite of the lower gross content of sodium ions in glass No. 4 than in glass No. 5, the conductivity of the former overtakes the conductivity of the latter at a sufficiently high temperature as a result of a more rapid increase in the degree of dissociation with temperature caused by a more pronounced decrease in the free energy of dissociation $\Delta\Phi$ and in the associated Ψ_Φ parameter.

It is known that an increase in the temperature corresponding to the lower limit of the critical region T_0 with an increase in the alkali oxide content of the glass is caused by a strengthening of the covalent structure. The latter occurs by virtue of the increase in the number of linking valence bonds in the system [7]. This may explain the increased entropy effect in the more friable glass No. 4 compared with No. 5 (assuming that the entropy effects ΔS are approximately the same in the temperature range in which we investigated the conductivity shown in Fig. 1).

It follows that the cross–over of the conductivity curves is specific to borate glasses and may be explained by the strengthening of the structural chemical network of valence bonds

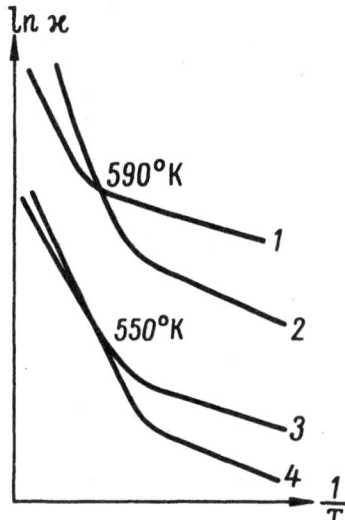

Fig. 4. Cross-over of conductivity curves for borate glasses with different alkali oxide content.

between the atoms on the transition from the trivalent neutral boron atom to the negatively ionized quadrivalent atom when the metal oxide content of the glass is increased.

In silicate, in contrast to borate glasses, a reverse pattern is observed. In this case, as the alkali oxide content increases the number of connecting valence bonds decreases and the lower limit of the critical temperature region T_0 falls. As a result, the conductivity curves (equivalent to those of Fig. 4) for silicate glasses in the labile state diverge rather than approach each other. In silicate glasses, therefore, the cross-over of the conductivity curves is not observed.

7. In glass of a low alkali oxide content the dielectric loss is caused basically by activated ion polarization. This must be established considerably more rapidly in the labile than in the stable state [10] and the phenomenon of the decrease in initial current with time normally disappears in the labile state. It is possible, however, that in the critical region there is superposition of the initial absorption current and of the conductivity current which is rapidly increasing with temperature due to the entropy effect considered earlier. This must clearly explain the fact that the dielectric loss of glasses at power frequencies and high temperatures consists almost completely of conductivity loss [4, p. 420].

According to Skanavi, in borate glasses above a certain temperature, the dielectric loss, tan δ, begins to increase very rapidly. The lower the alkali oxide the higher is this temperature (in glass No. 14). This temperature of the initial rise in the loss does not have anything in common with the lower temperature of the critical range. In borates the latter falls with a decrease in the alkali oxide content [7]. In the phenomenon noted by Skanavi there is certainly an overlap of the conductivity loss by a constant loss which is independent of temperature. This overlap is also produced by a shift in the temperature of initial growth towards the lower temperatures and to the high alkali oxide content glass.

This overlap can therefore explain the appearance of the temperature dependent tan δ in boric oxide at 350°C. The critical temperature in boric oxide, about 250°C, is considerably lower. When the glass is quenched the increased loss tangent is consistent with the increase of its electrical conductivity. We would expect a decrease in Ψ_Φ due to the partial retention of the high temperature geometrically disordered packing of the polar structural groups. Such disorder must be accompanied by some decrease in the bond energy of the cations associated with polar groups and thus produce an increased degree of dissociation.

Therefore, with the transition to complete geometrical order in crystalline silicates the degree of dissociation must decrease a little and the conductivity must fall. Botvinkin, Okhotin and other workers have established the latter in [12]. Results presented by Schwartz and Gal'bershtadt [7], according to which a slight increase in the temperature coefficient of the electrical conductivity is observed on the transition from vitreous to crystalline sodium disilicate, indicate that there is such a small increase in the energy Ψ_Φ.

The decrease in tan δ observed on crystallization must also be due to the strengthening of the cation bonds in the geometrically ordered lattices and not to the "close packing of molecules" [11]. It is sufficient to point out that the dielectric loss decreases significantly on crystallization (by 50 to 90%) while the density changes only slightly. Incidentally, cases of a decrease in the density of glass on crystallization (for example, borax from 2.37 in glass to 2.28 in the crystal [13]) may be noted.

It is still not quite clear to what extent the existence of dislocations in the form of breaks in the continuity of the material, occurring, it would seem, in polycrystalline structures, contributes to the decrease in conductivity on crystallization. It is possible that this kind of effect is attenuated to a considerable extent by the mutual interconnection of the crystallites by transition zones comprising semiordered arrangements of atoms [14]. We cannot ignore the facts which definitely point to the existence in crystals of such a "mechanical" factor which decreases electrical conductivity.

Thus Moriyasu [15] established that feldspar, both in the crystalline and in the vitreous state, has a single energy coefficient, A of Eq. (1), of 10,600, while the B coefficient, normally + 4.8 in glasses, falls after crystallization to −6.0. These results definitely seem to indicate the absence, on the crystallization of feldspar, of a considerable change in the dissociation energy and to the high significance of a mechanical factor such as breaks in the continuity of the material with a polycrystalline structure.

Conclusions

1. Three methods of determining the value of the characteristic parameters of electrical conductivity of glasses in the labile state have been critically examined: Ψ_Φ, heat Ψ_H and entropy ΔS parameters.

2. Values have been given for $\Psi_\Phi(T)$, $\Psi_H(T)$, and $\Delta S(T)$ for borates and silicates in the temperature region in question.

3. In glasses in the labile state the parameters $\Psi_\Phi(T)$, $\Psi_H(T)$, and $\Delta S(T)$ are functions of temperature; Ψ_Φ invariably decreases as the temperature increases thus producing an increased rate of rise of conductivity with temperature.

4. In the critical temperature region electrolytic dissociation is accompanied by an entropy change which produces a maximum in the $\Delta S(T)$ curve. The ascending section corresponds to the growth, at the beginning of the critical region, of local structural charges on dissociation (activation of valence oscillations). The descending section of the $\Delta S(T)$ curve corresponds to the disappearance of local structural changes on dissociation close to the melting temperature (limited destruction of the geometrically ordered packing of structural elements as a result of the intensive disordered thermal vibrations of the atoms).

5. The temperature dependence of the heat parameter $\Psi_H(T)$ is determined basically by the form of the change in the entropy of dissociation $\Delta S(T)$.

6. A physical model of the nature of the changes of the characteristic parameters in glasses with temperatures in the labile state has been presented.

7. The phenomenon of the cross-over of the curves of electrical conductivity against T in borate glasses was elucidated and the changes in the conductivity of the glasses on quenching and on crystallization were briefly considered.

8. It was recommended that, in the publication of experimental research on the conductivity of glasses, the complete numerical experimental results for the specific conductivity at various temperatures should be recorded. These results are essential for strictly valid calculations of the characteristic parameters Ψ_Φ, Ψ_H, and ΔS.

References

1. R. L. Myuller, Zh. Tekhn. Fiz., 25:246 (1955), ● this volume, p. 24.
2. S. A. Shchukarev and R. L. Myuller, Zh. Fiz. Khim. 1:625 (1930); S. A. Shchukarev (Schtschukarew) and R. L. Myuller (Müller), Z. Phys. Chem., A150:439 (1930); R. L. Myuller and B. I. Markin, Zh. Fiz. Khim., 5:1272 (1934); R. L. Myuller (Müller) and B. I. Markin, Acta Physiocochim. URSS, 1:266 (1934); 4:471 (1936); R. L. Myuller and B. I. Markin, Zh. Fiz. Khim, 7:592 (1936); B. I. Markin, Zh. Obshch. Khim, 11:285 (1941); L. P. Takking and N. P. Shchegoleva, Uch. Zap. Leningr. Gos. Univ. No. 108, p. 17 (1949).
3. P. P. Kobeko and N. I. Shishkin, Zh. Tekhn. Fiz., 17:27 (1947).
4. G. I. Skanavi, Dielectric Physics GTTI, Moscow (1949).
5. S. Oka, J. Soc. Chem. Ind. Japan, 30:625 (1927); C. A., 22:1525 (1928); Gmelin's Handbuch der anorganischen Chemie, Syst-Numm 21, 8 Aufl., (1928), S. 657.
6. R. L. Myuller, Uch. Zap. Leningr. Gos. Univ., No. 54, p. 159 (1940), ● this volume, p. 3.
7. R. L. Myuller, The Vitreous State and the Electrochemistry of Glass (Doctoral Dissertation), Leningr. Gos. Univ. (1940); Zh. Fiz. Khim, 28:2170 (1954).
8. R. L. Myuller, Zh. Fiz. Khim., 6:616 (1935); R. L. Myuller (Müller), Acta Physiocochim. URSS, 2:103 (1935).
9. E. Seddon, E. Tippett, W. E. S. Turner, J. Soc. Glass Technol., 16:459, (1932).
10. R. L. Myuller, Zh. Tekhn. Fiz., 25:1556, 1557 (1955), ● this volume, pp. 33, 43 respectively.
11. G. I. Skanavi, Zh. Tekhn. Fiz., 7:1039 (1937); Élektrichestvo, No. 8, 15 (1947).
12. K. S. Evstrop'ev and N. A. Toropov, Silicon Chemistry and the Physical Chemistry of Silicates, Promstroiizdat, Moscow (1950), p. 114; A. M. Samarini and L. A. Shwarzman, Usp. Khim., 21:336 (1952); M. Fulda, Sprechsal Keramik, Glas, Email, 60:769, 789, 810, 831, 853 (1927); E. M. Guyer, J. Amer. Ceram. Soc., 16:607 (1933).
13. Gmelins Handbuch der anorganischen Chemie, Syst.-Numm., 21, Natrium, 8 Aufl. (1928), p. 656.
14. V. I. Arkharov, Zh. Tekhn. Fiz., 22:332 (1952); M. V. Klassen-Neklyudova, and T. A. Kontorova, Usp. Fiz. Nauk, 22:249, 395 (1939).
15. J. Moriyasu, Japan Ceram. Assoc., 45:149 (1937); G. Z., 1:3924 (1937).

VII

The Temperature Dependence of the Electrical Conductivity of Crystalline Materials*

1. A typical feature of ionic-covalent high melting substances is the existence of a low temperature stable state in which the degrees of freedom of the valence oscillations of the atoms are frozen [1]. This feature predetermines a series of properties in these substances, in particular the presence of "anentropic" electrical conductivity in the stable state changing to "entropic" in the labile state. The "entropy effect" in conductivity is explained by the excitation of valence oscillations and by the associated local changes in the degree of geometrical order in the structure which in the labile state accompanies the electrolytic dissociation of the ionic-covalent systems [2].

Experimental and theoretical studies of conductivity in ionic-covalent substances have been made on such substances in the vitreous state [3]. Since the temperature of transition, however, from anentropic to entropic conductivity is associated with the transition of the particular substances from the stable to the labile state and results from the covalent nature of the chemical structure, it can be stated that these electrochemical features are characteristic not only of the vitreous but also the crystalline state of such substances. In fact, as the thermal capacity data suggest, the chemical structure of ionic-covalent substances does not depend on purely geometrical features of the crystalline lattice nor on the existence of an ordered crystalline or disordered vitreous arrangement in these substances. The phenomenon of strict valence bond stabilization at low temperatures is common to such substances both in the vitreous and crystalline state [1, 4]. We should thus expect that the results obtained from a study of the conductivity of vitreous borosilicates would, in the main, also characterize the electrical conductivity of crystalline borosilicates. This will be true, primarily, of the temperature dependence of the electrical conductivity.

When borosilicates change from the vitreous to the crystalline state there is some decrease in their conductivity. It is probably due to the increase in dissociation energy produced by the increased strength of the cation bonds when the geometrically ordered lattice is established. The decrease in conductivity may also be caused by the breakdown of the continuity of the material on the transition to the crystalline structure [3].

There is at present too little experimental material on the electrical conductivity of crystalline borosilicates to check these ideas.

2. At present, data on the electrical conductivity of the simplest ionic crystals is more extensive and firmly based. The analysis of the temperature dependence of the conductivity of such crystalline bodies is particularly interesting in connection with the results of the study of the conductivity of ionic-covalent systems.

*R. L. Myuller, Zh. Tekhn. Fiz., 25(14):2440 (1955).

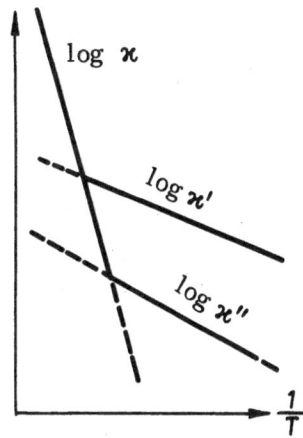

Fig. 1. High temperature ($\varkappa$) and low temperature ($\varkappa'$, $\varkappa''$) electrical conductivity in sodium chloride

It is well known that the temperature dependence of the electrical conductivity is the same for all alkali halide crystals. It is expressed by the empirical relationship

$$\varkappa = C \exp\left(-\frac{A}{T}\right),\tag{1}$$

giving a log-linear dependence of conductivity on the reciprocal of the absolute temperature. Taking into account the identity of the temperature dependence of conductivity of all alkali halide crystals we can restrict ourselves to a detailed analysis of any one halide, such as sodium chloride.

3. In crystalline sodium chloride the high and low temperature conductivities are distinct. Each of them shows a log-linear dependence but they differ in the slopes of the lines (see Fig. 1). The electrical conductivity observed at high temperatures ($\varkappa$) is satisfactorily reproducible in different specimens of crystalline sodium chloride.

At low temperatures the slope of the lines is considerably less than at high temperatures. The lack of reproducibility of this value for different specimens of sodium chloride is a feature of the low temperature electrical conductivity. An example of this is shown in Fig. 1 by the curves of low temperature conductivity (log $\varkappa'$ and log $\varkappa''$) experimentally observed in two specimens.

There is no doubt at all that the relatively high temperature electrical conductivity is due to the movement of the ions of the crystalline lattice, the sodium ions. It has been considered that only high temperature conductivity is a characteristic of a crystal [5, pp. 253, 267, 268, and 275].

The nature of the low temperature conductivity of ionic crystals has not yet been very convincingly explained. It is a widely held view that the low temperature conductivity is poorly reproducible as a result of its obvious dependence on all types of extrinsic ionic impurities in the crystal [5, pp. 268, 275]. Low temperature conductivity is attributed to ions in defect sites in the crystalline lattice [5, p. 253]. This does not exclude the additional factor of the movement of ions along the cracks in the polycrystalline structure, the "structure sensitive conductivity" [6, pp. 52, 57].

It is unlikely that low temperature conductivity is conditional on the presence of impurity ions. In non-ionic dielectrics like boron or silicon oxide, the intrinsic conductivity is so small that even quite insignificant contamination by ionic inclusions can, in fact, determine completely the conductivity observed in them. But in halides consisting entirely of alkali metal ions the role of impurity ions in electrical conductivity is doubtful. The relative content of the ionic impurity is so insignificant that even allowing that all the impurity ions are freely mobile, it cannot account for the increased value of the electrical conductivity at low temperatures.

On the other hand we can present a convincing argument in favor of the idea that low temperature conductivity in sodium chloride is caused, like the high temperature conductivity, by the intrinsic sodium ions. Nevertheless, the nature of the intrinsic conductivity at low temperatures must be essentially different from the nature of the high temperature conductivity.

4. Primarily we should consider the quantitative aspect of the principle which determines the relative positions of the conductivity curves in Fig. 1. From the experimental data reproduced graphically by Skanavi in [5, pp. 273, Fig. 82] we can consider as experimentally established at least three segments of the conductivity curves with values of C_e and A determined from Eq. (1) and shown in Table 1.

TABLE 1. Crystalline Sodium Chloride

Electrical conductivity	T, °K	C_e	A	$\log \dfrac{C_e}{C_t}$	ΔS, kcal/mole · deg.	Ψ_H, kcal/mole	Ψ_Φ, kcal/mole		
							723	973	1073
High temperature ($\varkappa$)	>723	$1.0 \cdot 10^6$	22200	4.7 ± 0.5	43 ± 5	88.0	56.9	46.2	41.8
Low temperature									
$\varkappa'$	<973	40	11500	$+0.3$	—	—	46.2	46.2	—
$\varkappa''$	>723	12.6	13600	-0.2	—	—	54.0	—	—

For the case of conductivity due to the movement of the basic ions of the crystalline lattice Frenkel' theoretically deduced an expression analogous in form to the experimental expression, Eq. (1). According to Frenkel's theory [1, 7, and 8] the value of the pre-exponential term for any crystalline body can be calculated from the formula

$$C_t = 3 \cdot 10^{-21} z^2 n_0 \, \Omega^{-1} \cdot \text{cm}^{-1}, \tag{2}$$

where z is the valence of the current-carrying ions, and n_0 is the concentration of these ions in the crystal expressed in cm^{-3}. In this case of sodium conductivity, z is 1 and the concentration in the sodium chloride is $2.2 \cdot 10^{22} \, \text{cm}^{-3}$. Substituting these values in Eq. (2) and remembering that in obtaining the latter the oscillation frequency of the ions was assumed to be $\nu = 10^{13} \, \text{sec}^{-1}$, but may in fact be as low as 10^{12}, we obtain the numerical limits for the parameter $\log C_t$:

$$0.82 \leqslant \log C_t \leqslant 1.82,$$

or

$$\log C_t = 1.3 \pm 0.5.$$

A comparison of the theoretical value C_t with the experimental value C_e, expressed as $\log C_e/C_t$, establishes the important fact that there is extremely good agreement between the experimental and theoretical values to a high degree of accuracy, for both curves for the low temperature conductivity of sodium chloride (see Table 1). At the same time we must note that the experimental value of the pre-exponential term is $5 \cdot 10^4$ greater than the theoretically determined value in the case of the high temperature electrical conductivity.

There is a direct proof that the low temperature conductivity in sodium chloride is determined by the basic ions of the crystalline lattice. (Their number was used in Frenkel's calculation). We know that at low temperatures the mobile ions are sodium ions. At the same time, equating the $\log C_e/C_t$ to zero, in fact, is equivalent to equating the entropy of electrolytic dissociation to zero

$$\Delta S = 9.15 \, (P_e - P_t) = 9.15 \log \frac{C_e}{C_t} \; \text{cal/mole} \cdot \text{deg.} \, , \tag{3}$$

since from the relationship previously established $P_e = \log C_e / [M]$ and $P_t = \log C_t / [M]$.

Thus, the numerical values of the preexponential term C_e in the expression for the temperature dependence of the low temperature conductivity of various specimens of sodium chloride crystals points to the likelihood that this conductivity is of an anentropic character and is due to mobile intrinsic sodium ions. In such a case the energy parameter Ψ_Φ^0 is independent of temperature and is given by $\Psi_\Phi = 3.97$ A kcal/mole [2]. The value of this parameter for the low temperature conductivity obtained in this way are given in Table 1.

5. The considerable excess of the experimental value of the pre-exponential term over the theoretically calculated value for the high temperature conductivity indicates that the electrolytic dissociation which determines this conductivity is accompanied by an entropy effect. The approximate value of the entropy ΔS from Eq. (3) is shown in Table 1. The values of the heat parameter $\Psi_H = 2AR$ and the free energy parameter $\Psi_\Phi(T) = \Psi_H - T\Delta S$ for three temperatures (in °K) [2] are also given in the table.

The entropy change on electrolytic dissociation is quite reasonable in the case of purely ionic solids. The heat capacity of such solids under normal conditions points to the absence among the ions of the frozen degrees of freedom of thermal vibrations [4]. Thus according to Frenkel' [7], with the displacement of a cation from a normal lattice point into an irregular interstitial position, local breakdowns must occur in the geometrically ordered arrangements of the neighboring cations at the lattice points [2]. Thus a local breakdown in order will be observed around a dissociated cation in an interstitial position, and no doubt to smaller extent, around a vacant site at a lattice point.

The mechanism itself of current flow through a crystal can be reduced to the movement of vacancy sites as a result of the direct displacement of a cation from regular positions into the neighboring vacancy sites. But the number of such vacant sites producing the high temperature conductivity must, according to Frenkel', be determined by the number of dissociated interstitial cations. Such electrolytic dissociation in the labile medium of a purely ionic crystal will be accompanied by local changes in entropy which will produce the dependence of the energy parameter $\Psi_\Phi(T)$ on temperature. According to the data in Table 1, $\Psi_\Phi(T)$ falls with an increase in temperature producing a greater increase in the electrical conductivity with temperature which explains the steeper slope of the line representing log $\varkappa$ in Fig. 1. It is well known that this slope roughly characterizes the increased heat parameter Ψ_Φ. The increased value of the pre-exponential term is due to the presence of an additional entropy term.

6. If ionic crystalline bodies are characterized by a labile structure and the resulting entropic character of the electrical conductivity, then what produces the lack of an entropy effect in the low temperature conductivity? In solving this problem we have to take as proved the fact that at low temperatures, as also at high temperatures, electrical conductivity is produced by the intrinsic sodium ions (the theoretical basis for the experimental value of the pre-exponential term in sodium chloride).

It has been shown previously that the stabilization of the structure by the establishment under normal conditions of some rigid bonds between ions must, in principle, be excluded in sodium chloride. From an atomic-kinetic point of view this embraces the fundamental distinctions between purely ionic substances and ionic-covalent; between sodium chloride and borosilicates. We should therefore expect the nature of the anentropic low temperature conductivity in sodium chloride to be radically different from the nature of such conductivity in borosilicates.

It is known that, as well as electrolytic dissociation of the Frenkel' type, the development of vacancy sites at the lattice points of ionic crystal ("holes") can arise by the transport of an equal number of ions of both signs from the internal region of the crystal to the surface where they are lost (Schottky's mechanism [9]). Such a mechanism for the creation of vacant sites is only possible in ionic crystals and is impossible in borosilicates [1].

The absence of disordered cation penetration into irregular interstitial positions is char-acteristic of the Schottky mechanism of obtaining vacant sites. Meanwhile, as we have seen earlier, the entropy effect in electrical conductivity is most probably determined by this type of disordered penetration into the interstitial positions. The appearance of vacancy sites at regular lattice points in itself hardly initiates any change in the entropy of the system.

If this is correct we must then recognise that an entropy effect is not always necessary for electrical conductivity in ionic crystals. The entropy effect must be observed when vacancy sites appear at the lattice points according to the Frenkel' mechanism and must, in fact, be ab-sent when Schottky vacancy sites are formed. Assuming that this same mechanism (the dis-placement of vacancy sites at lattice points) is unchanged at all temperatures then we have to distinguish the high temperature mechanism for the development of vacancy sites according to Frenkel', from the low temperature mechanism according to Schottky. In making such a demar-cation we can understand the presence of an entropy effect in the well reproducible high temper-ature conductivity and the anentropic character of the poorly reproducible low temperature con-ductivity.

In fact, the mechanism of generation of vacancy sites by displacement of cations into in-terstitial positions is accompanied by the entropy effect and is reproduced in any specimen of sodium chloride crystals in accordance with the normal laws of statistical physics.

The mechanism for the appearance of vacancy sites by the discharge of an equivalent num-ber of positive ions from the volume of the crystal at the surface is not, as was shown previous-ly, accompanied by a noticeable change in entropy. The energy of such a transition will also de-pend on the dimensions of the crystallites of which the polycrystalline sodium chloride consists. Bearing in mind the Madelung coefficient we can see that with a decrease in the dispersivity of the polycrystalline structure the energy of development of the vacancy sites will increase slight-ly and the electrical conductivity will decrease.

7. A critical analysis of the values of the coefficients in the conductivity equation (Eq.(1)) and the subsequent discussion leads to the following explanation of the observed features of the electrical conductivity of crystalline sodium chloride.

The conductivity is realized by the interchange of the vacancy sites at the lattice points with an activation energy E_A. The exponential rise of conductivity with temperature is produced primarily by the exponential rise in the concentration of vacancy sites at the lattice points.

Vacancy sites develop through two mechanisms: 1) by the discharge of cations from their normal position at lattice points into irregular interstitial positions. This transition is accom-panied by an entropy change and the free energy of the transition consequently drops noticeably with an increase in temperature; 2) by the transition of an equivalent number of ions of both signs from the volume to the surface of the crystallites in the polycrystalline structure. This transition is not accompanied by any marked local entropy changes and the transition energy in fact does not depend on temperature. The transition energy increases slightly with a decrease in the dispersivity of the polycrystalline structure.

At low temperatures the energy of the entropic generation of vacancy sites is higher than the energy of the anentropic process. At high temperatures the decreasing energy of the entrop-ic generation becomes less than the constant energy of the anentropic formation of vacancy sites. As a result, the observed electrical conductivity of sodium chloride

$$ \varkappa = C\left[\exp\left(-\frac{\Psi_\Phi(T)}{2RT}\right) + \exp\left(-\frac{\Psi_\Phi^0}{2RT}\right) \right] $$

at high temperatures is determined by the overwhelming number of vacancy sites developing with an entropy change. The corresponding energy parameter of the conductivity satisfies the inequality $\Psi_\Phi(T) < \Psi_\Phi^0$.

The low temperature conductivity is determined essentially by the number of vacancy sites generated without an entropy change. The corresponding energy parameter of the conductivity satisfies the inequality $\Psi_\Phi^0 < \Psi_\Phi(T)$ as a result of the increase in the value of Ψ (T) with a decrease in temperature while Ψ_Φ^0 is invariant.

The low temperature electrical conductivity is poorly reproduced in different crystal specimens due to the dependence of the Ψ_Φ^0 parameter on the dispersivity of the polycrystalline structure of the sodium chloride.

The pre-exponential statistical factor C is retained without change at all temperatures under conditions when the entropy term is included in the exponential function.

Everything above which applies to the particular case of the electrical conductivity of sodium chloride can be applied to the conductivity of any ionic crystals.

We should expect that the low temperature conductivity of single crystals of an alkali metal halide would be characterized by good reproducibility and that the value of Ψ_Φ^0 will be constant and higher than the value of that parameter in polycrystalline specimens of the same halide.

Conclusions

1. In crystalline, as in vitreous borosilicates, we should expect a transition from the anentropic conductivity of the stable state at low temperatures to the entropic conductivity of the labile state in the critical temperature region.

2. An experimental and theoretical analysis of the temperature dependence of the conductivity of sodium chloride leads to an understanding of the nature of this dependence in purely ionic crystalline bodies.

a) High temperature and low temperature conductivity are due to the movement of the intrinsic ions of the crystalline lattice.

b) High temperature electrical conductivity is determined by the number of vacancy sites at lattice points, which develop according to Frenkel' as a result of the discharge of ions into interstitial positions. This discharge is accompanied by an entropy change.

c) Low temperature electrical conductivity is determined by the number of vacancy sites, which according to Schottky are generated by the discharge of ions at the surface of the crystallites of the polycrystalline structure without an entropy change.

d) The energy of the entropic generation of vacancy sites $\Psi_\Phi(T)$ decreases when the temperature is increased. At low temperatures its value exceeds the value of the energy of the temperature independent anentropic generation of vacancy sites Ψ_Φ^0. At high temperatures the inequality is reversed, $\Psi_\Phi(T) < \Psi_\Phi^0$.

e) The poor reproducibility of the low temperature conductivity is explained by the dependence of the energy of the Schottky type of vacancy-site generation on the dispersivity of the particular polycrystalline structured specimen of the ionic substance.

References

1. R. L. Myuller, Zh. Tekhn. Fiz., 25:236 (1955), ● this volume, p. 15; Zh. Fiz. Khim., 28:1193, 1831 (1954), 2170 (1954).
2. R. L. Myuller, Zh. Tekhn. Fiz., 25:246 (1955), ● this volume, p. 24.
3. R. L. Myuller, Zh. Tekhn. Fiz.,25:1700 (1955).
4. R. L. Myuller, The Vitreous State and the Electrochemistry of Glass (Doctoral Dissertation), Leningr. Gos. Univ. (1940); Zh. Fiz. Khim., 28:1193 (1954).
5. G. I. Skanavi, Dielectric Physics, GITTL, Moscow (1949).

6.   N. Mott and R. Gurney, Electron Processes in Ionic Crystals, Oxford Clarendon Press (1940).
7.   Ya. I. Frenkel', Z. Phys., 35:652 (1926).
8.   Ya. I. Frenkel', A Kinetic Theory of Liquids, Izd. Akad. Nauk SSSR, Moscow (1945).
9.   C. Wagner and W. Schottky, Z. Phys. Chem., B11:163 (1930).

VIII

The Concentration Dependence of the Electrical Conductivity of Borate and Silicate Glasses*

1. The electrical conductivity of glasses is directly dependent on their volume concentration of ionic oxides. The composition determines the structural chemical features of the vitreous state which in its turn has a profound effect on the electrical conductivity of the glass. The electrical conductivity of glass has thus a unique functional dependence on the volume concentration of the ionic structural groups.

The absence of well-defined structural chemical features of the vitreous state is a shortcoming in some of the proposed statistical interpretations of the low temperature conductivity of glasses [1]. In the analysis of electrolytic dissociation in thermodynamic equilibrium no allowance was made for the thermodynamically nonequilibrium state of the glass structure as a whole. Moreover, in the critical temperature region, structural processes occuring on electrolytic dissociation were underestimated. To date, therefore, it was not possible to elucidate the physical nature of the observed concentration dependence of conductivity in glasses.

2. The most extensive experimental study has been our work on the concentration dependence of the conductivity of vitreous alkali borosilicates. The similarity in the character of this dependence for all the alkali metals has been established [2]. We can therefore confine ourselves to a detailed consideration of the experimental data for a single system, such as the sodium borate glasses [3]. Henceforth, we shall also make use of the experimental and theoretical data for the energy parameter Ψ_Φ which we established in previous communications [4, 5].

Figure 1 shows the experimental data for sodium borate glasses. The log molar conductivity is shown as a function of the log concentration of sodium ions, expressed in mole/ ml. The energy of conductivity Ψ_Φ, in kcal/mole, is also plotted on the abscissa. The results refer to two temperatures: 200 and 300°C.

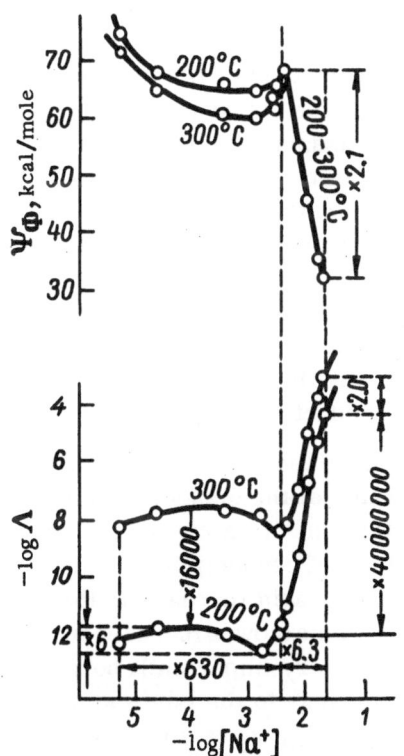

Fig. 1. The dependence of the energy Ψ_Φ, and the log molar conductivity of borate glasses on log sodium ion concentration.

*R. L. Myuller, Zh. Tekhn. Fiz., 26(12):2614 (1956).

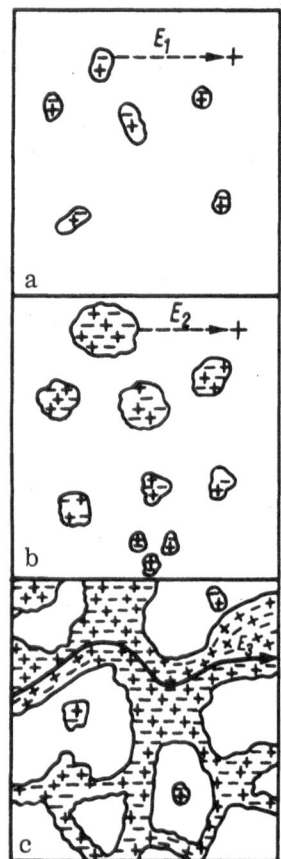

Fig. 2. The growth of associated groups of polar structural units with an increase in their concentration in the glass: a) relatively low content of polar structural units; b) increased number and size of associated groups; c) coalescence of units with the emergence of continuous conductivity in the polar medium.

The parameter Ψ_Φ, called from now on, for the sake of brevity, the "dissociation energy" is an approximate measure of the free energy of electrolytic dissociation [6]. It is clear from Fig. 1 that the initial 630-fold rise in the sodium ion concentration in the glass is accompanied by a minor fluctuation in the molar conductivity, with over all not more than a 6-fold change. A further increase, by only six times in all, of the sodium content is accompanied by a $4 \cdot 10^7$ jump in the molar conductivity, consistent with the known formula

$$\Lambda = C \exp\left(-\frac{\Psi_\Phi}{2RT}\right). \tag{1}$$

The concentration dependence of molar conductivity must be basically determined by the dissociation energy Ψ_Φ [4-6]. Accordingly there are two clearly delineated segments in the dissociation energy curves of Fig. 1 corresponding to the following regions of the sodium concentration in the glass: first, a low concentration region which is characterized by a relatively small change in the energy Ψ_Φ; and second, a concentration region which is associated with a rapid twofold decrease in the energy.

In the first region the dissociation energy first decreases slightly and then begins to increase. Correspondingly, the molar conductivity begins to increase slightly at first and then decreases. The first decrease in the dissociation energy is explained by an increase in the dielectric constant as the polar structural unit concentration increases (the solvation effect [7]). The subsequent slight increase in Ψ_Φ is the result of the stratified association of the polar structural units with the development of amicronodispersed polar inclusions which increase in number and size with an increasing Madelung coefficient. This results in an increase in the dissociation energy which is determined fundamentally by the work of release of the cation from the polar group into the ground medium of nonpolar structural units. In the second, narrow region of increased concentration, there is an abrupt drop in the dissociation energy which causes an exponential increase in the molar conductivity [4].

The size of this increase is noticeably less at 300°C than at 200°C. This is explained by the higher value of the dissociation energy in the region of low sodium concentration. As a result, the correspondingly higher temperature coefficient of conductivity in this region produces in the second region a more rapid increase in the lower electrical conductivity compared with the slower increase in the higher conductivity. Moreover, (and this is very important), borate glasses with a lower alkali concentration change to the labile state when the temperature reaches 300°C and as a result the dissociation energy Ψ_Φ is somewhat decreased (the entropy effect). Taken as a whole, the result is that the electrical conductivity of glasses depleted in alkali increases 16,000 times with a 100° rise in temperature, while the same increase in temperature increases the conductivity of vitreous borax (sodium borate) only 20 times.

Thus, with an increase in temperature a jump in the molar conductivity will occur. This conductivity jump which was subsequently confirmed by Spaght and Clark [8] for the same temperature region must, according to our forecast [9], disappear at temperatures above 500°C

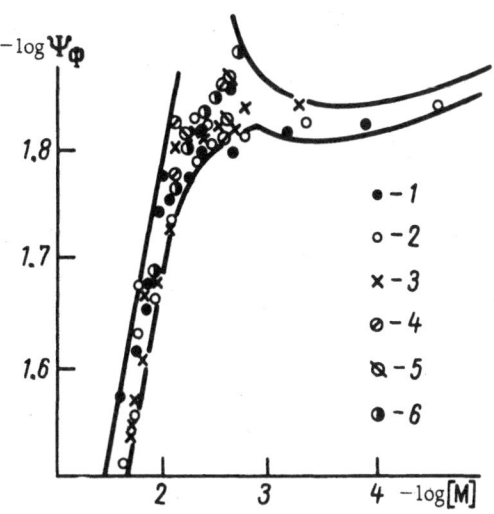

Fig. 3. Dependence of log Ψ_Φ on log ion concentration in borate glasses (stable state, 200°C): 1) lithium; 2) sodium; 3) potassium; 4) rubidium; 5) cesium; 6) silver.

when the concentration curve assumes a uniform, gentle slope. In fact, Stemaier and Dietzel [10] did not observe the jump of the molar conductivity in the systems at 700°C.

We noted in [4] that the clear transition from the moderate change in dissociation energy to the steep drop occurs at the critical concentration of sodium at which the growing but isolated polar microgroups (Fig. 2,b) come into contact and coalesce into polar microwebs (Fig. 2,b). Then, the conductivity which was limited by the movement of dissociated ions in the ground medium of the nonpolar structural units changes to conductivity determined by the movement of vacancy sites in the ground medium of polar structural units [11].

The numerical values of Ψ_Φ established in the author's recent studies [4, 5] confirm the general pattern of the nature of the concentration and temperature dependence of conductivity for all alkali glasses.

The change in the nature of the conductivity in the critical concentration region is a direct result of the change in the quantitative ratio of the number of nonpolar structural units [ζ] to the number of polar groups [ζ^-] [4, 6], thus

$$\gamma = \frac{[\zeta]}{[\zeta^-]} . \tag{2}$$

Here the concentration of polar units [ζ^-] corresponds to the content of structural units of the $B^-O_{4/2}M^+$ and $SiO_{3/2}O^-M^+$ type while that of the nonpolar units [ζ] corresponds to the $BO_{3/2}$ and $SiO_{4/2}$ types.

The transformation of the nature of the conductivity occurs when $\gamma \simeq 6$ [4]. In this region of concentration in the alkali borate glasses we noted the electrochemical series of decreasing mobilities [2]

$$\Lambda_{Li} > \Lambda_{Na} > \Lambda_K > \Lambda_{Rb}. \tag{3}$$

The dielectric loss in glasses follows the same series. In good agreement with this, when $\gamma \simeq 6$, sodium (No. 8), potassium (No. 52) and rubidium (No. 60) borate glasses have increasing dissociation energies Ψ_Φ: 55.1-60.0-67 kcal/mole [4, 5]. Such a series is the result of the increase in the cation radii and the increase in the corresponding dipole moments in the $B^-O_{4/2}M^+$ polar structural units, which determine the consequent rise in the degree of association and the increase in the Madelung coefficient in the resultant microdispersed groups. The latter explains the corresponding increase in the energy of cation escape from the associated polar groups in the nonpolar medium [7].

An exception is the lithium glass (No. 111) in which the value of $\Psi_\Phi = 56.5$ is higher than in the sodium glass. This may be explained by the increased polarization bond, noted earlier, of the lithium ion which is the smallest in mass and radius [12]. The simultaneous increase this causes in the vibrational frequency, which is reflected in the pre-exponential factor in Eq. (1) [6], compensates for the positive effect of the increased bond energy and thus the mobility of the lithium is maintained within the normal mobility range [3].

TABLE 1.

100 [Na], mole/ml	Ψ_Φ, kcal mole			100 [O], mole/ml			100 ([B] + [Si]), mole/ml			100 [...], mole/ml	
	borates	borosil-icates	silicates	borates	borosil-icates	silicates	borates	borosil-icates	silicates	borates	borosil-icates
0.6	62	54	37	7.8	8.0	7.2	5.0	4.8	3.5	15.6	13.8
2.3	32	32	30	8.1	8.1	6.9	4.6	4.3	2.9	16.2	11.5
Δ, %* ...	-48	-41	-23	+4	+1	-4	-8	-10	-23	+4	-17

* is obtained by subtracting the value in the first row from that in the second, and dividing the difference by the value in the first row.

With low alkali ion concentrations the electrochemcal mobility series can be considered as well established, since the difference in the numerical values of the molar conductivity reaches two orders of magnitude while there is a difference of 0.3 orders of magnitude between individual experimental values for different glass melts. For high alkali concentrations the differences in measured mobility lies within the limits of variation in the experimental measurements of other authors and it is therefore not possible to establish a mobility series in such circumstances.

3. In high alkali glasses we established a linear relationship between the log dissociation energy (in kcal/mole) and the log of the molar cation concentration (mole/ml) [9, 13]. Figure 3, based on our previous calculations in [4] and [5], shows this linear relationship.

In the case of the physical chemical interaction between ions and atoms in condensed systems, the linear variation in the dissociation energy in the narrow region of exponential increase in conductivity is of fundamental importance.

$$\log \Psi_\Phi = b - \varkappa \log[M]$$

or
$$\Psi_\Phi [M]^\varkappa = \text{const.} \tag{4}$$

In sodium borate glasses $\varkappa = 0.5$ and the constant is 3.5. In potassium borate and lithium borate glasses the constant is 3.9 [2]. Here, the empirical formula for the exponential dependence of molar conductivity on temperature (in °K) and cation content [M] (in mole / ml) is expressed in the form

$$\Lambda = C \exp\left(-\frac{a}{T\sqrt{[M]}}\right), \tag{5}$$

where $a = 930 \pm 50$ [14].

In sodium silicate glasses $\varkappa - 0.25$. This means that the dissociation energy in silicates decreases more slowly than in borates with an increase in the sodium concentration.

The nature of this difference between borates and silicates is complex. Some idea of the nature can be obtained using the data in Table 1. This table is based on the calculated data in [4] and [5]. The oxygen atom concentration was determined from

$$[O] = 1.5[B] + 0.5[Na] \quad \text{and} \quad [O] = 2[Si] + 0.5[Na].$$

Table 1 shows the closeness of the borosilicate to the borate data. This is explained, not only by the comparatively low proportional replacement of the boric oxide by silica

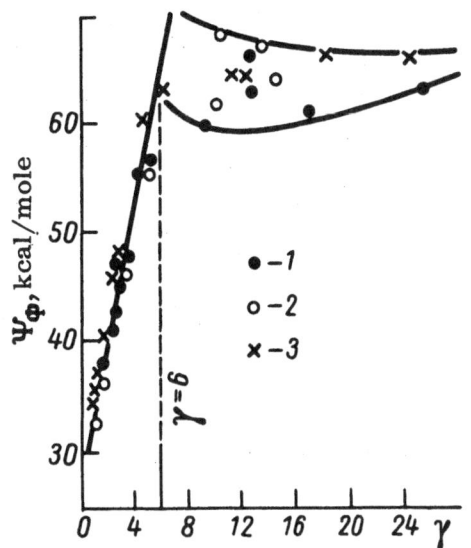

Fig. 4. Dependence of Ψ_Φ on the γ parameter in borate glasses (stable state, 200°C): 1) lithium; 2) sodium; 3) potassium.

(20% of the structural units in Markin's experiment [15]), but also by the predominant reaction of the sodium oxide with the boric oxide which is accompanied by an increase by one third in the number of valence bonds [16]. The concentration of linking valence bonds in sodium enriched silicate glasses ($[\ldots]_{Si} = 2[0] - [Na]$ mole/ml) is only 71% of the number of linking valence bonds in the corresponding borates ($[\ldots]_B = 2[0]$).

With a low sodium content ($6 \cdot 10^{-3}$ mole/ml) the solvation effect is essentially determined by the polarizability of the nonpolar boron and silicon oxides. The concentration of oxygen atoms in this case is lower in the silicate than in the borate (by 8%), since the concentration of silicon atoms is relatively lower than that of boron atoms (by 17%). The increased polarizability of silicon atoms, the radius of which (1.17 Å) is greater by 50% than that of the trivalent boron atom (0.78 Å), has a deciding effect in lowering the dissociation energy Ψ_Φ in silicates. The work of dipole destruction W_p must be higher in borates [7]. This follows from the values of the sum of the B⁻Na⁺ and O⁻Na⁺ radii (1.86 and 1.63 Å) which determine the dipole moments ($8.9 \cdot 10^{-18}$ and $7.8 \cdot 10^{-18}$ respectively) which are higher by 14% in borates.

An almost fourfold increase in the sodium content (up to $2.3 \cdot 10^{-2}$ mole/ml) produces a decrease in the dissociation energy Ψ_Φ (see Table 1) which is in borates twice that in silicates. This is explained by the large increase in the autosolvation effect in borates with an increase in the concentration of polar structural units. Secondly, the borate polar structural units have a markedly higher dipole moment. Thirdly, with an increased sodium content the concentration of oxygen atoms increases in borates while it decreases in silicates. It is seen in Table 1 that the relative decrease of the boron content in borates (−8%) is three times less than the silicon atom content in silicates (−23%). At the same time the radius of trivalent boron (0.78 Å) in borates increases by 12%, reaching the radius of quadrivalent boron (0.88 Å).

Qualitatively, therefore, the nature of the sharp decrease in the dissociation energy of borates is understood. The relatively smaller solvation effect in borates low in alkalis increases with an increase in the alkali oxide content considerably more than in silicates.

The quantitative aspect of Eq. (4) requires a further special study.

Equation (4) is from the physical point of view not entirely satisfactory. On changing from borates to silicates the dimension of the constant changes. The dependence of dissociation energy on the γ parameter discussed by the author [17] is more rational, as it is based on the calculations in [4] and [5]. This relationship is shown in Fig. 4.

In a medium of polar structural units the dissociation energy Ψ_Φ is linearly dependent on the γ parameter when $\gamma \leq 6$

$$\Psi_\Phi = \Psi_\Phi^0 + \beta\gamma, \tag{6}$$

where β and Ψ_Φ^0 are constants. It is evident here that $\Psi_\Phi^0 = \Psi_\Phi$ when $\gamma = 0$, i.e., according to Eq. (2) when $[\zeta] = 0$. This corresponds to the point at which single ionization of all the structural units has occurred and to the simultaneous disappearance of the nonpolar $[BO_{3/2}]$ and $[SiO_{4/2}]$ units.

TABLE 2

Substance	Ψ_Φ^0, kcal/mole	β, kcal/mole
Lithium borates	27.5	6.00
Sodium borates	27.0	5.50
Potassium borates	28.25	6.75
Sodium silicates.	30.0	1.88

A graphical analysis of the experimental data [4, 5] makes it possible to establish the values for the constants (when $\gamma \leq 6$), shown in Table 2.

Both in borates and in silicates in the stable state the dissociation energy approaches the value $\Psi_\Phi \approx 22$ kcal/mole when the single ionization of the structural vitreous units is completed. The generalized expression for molar conductivity, Eq. (5), can now be rewritten as

$$\Lambda = C \exp\left(-\frac{7000 + 0.25\beta\gamma}{T}\right), \tag{5a}$$

which is valid, in fact, for any alkali borosilicate.

4. When $\gamma = 0$ the structural units are found to be singly ionized and with the subsequent increase in the metal oxide content of the glass, double ionization occurs. The multiplicity of ionization can be increased. In these conditions we must use, as well as the concentration relationship γ, a parameter expresssing the multiplicity of the ionization of the structural units; single, α'; double, α'' [4]; triple, $\alpha''' = \dfrac{[\zeta\equiv]}{[\zeta\equiv] + [\zeta=]}$; and quadruple, $\alpha''' = \dfrac{[\zeta\equiv]}{[\zeta\equiv] + [\zeta\equiv]}$.

According to the previous paper [4],

$$\alpha''' = \frac{z\,[M]}{[X]} - 2; \quad \alpha'''' = \frac{z\,[M]}{[X]} - 3, \tag{7}$$

where [X] is the boron or silicon atom concentration in mole/ml. For quadruple ionization of the silicate units an isolated complex $[SiO_4^-]^{4-}$ ion is found, not connected by valence bonds to other structural units. The quadruply ionized $[B^-O_3^-O_{1/2}]^{4-}$ borate structural unit maintains one valence bond with a neighboring structural unit. The extra valence bond, for an equal multiplicity of ionization in borate structural units compared with silica units, was noted previously in [4].

Figure 5 shows the dependence of the dissociation energy of borates of univalent metals on the degree of the single ionization of the structural boric oxide units. The values of the dissociation energy of sodium in the presence of divalent barium $[Ba^{2+}] = 0.0082$ mole/ml are also given [4, 5]. It is known that alkali earth ions not only do not produce appreciable conductivity by themselves, but in fact decrease noticeably the conductivity of univalent cations such as sodium ions [18]. The dissociation energy increases significantly in this case (see Fig. 5) while the C factor in Eq. (1) remains unchanged [4]. As a result of the impossibility of orientation displacements of the barium ions in the medium of polar barium structures when the concentrations are not very large ($\alpha' \leq 1$), the corresponding solvation effect is absent [11] and in particular, therefore, the dielectric loss is increased only very slightly by the barium ions as indicated by the results of Val'ter et al. [19]. For the same reason the potential energy of the dissociated univalent ion in a medium of polar divalent metal borates increases relative to the potential energy of the same dissociated ion in a univalent metal borate medium [Fig. 6]. The dissociation energy increases correspondingly from 40–45 to 55–60 kcal/mole (Fig. 5). The

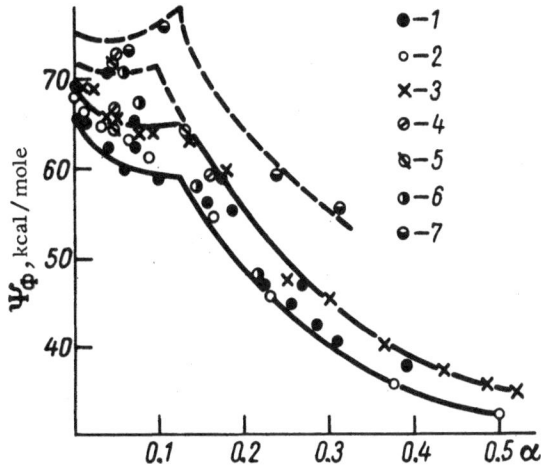

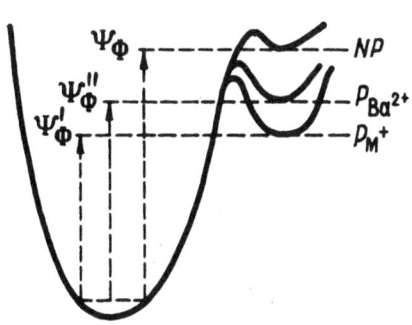

Fig. 5. The dependence of the energy Ψ_Φ on the degree of ionization α' in borate glasses, (stable state, 200°C): 1) lithium; 2) sodium; 3) potassium; 4) rubidium; 5) cesium; 6) silver; 7) sodium bariate.

Fig. 6. The energy Ψ_Φ for the electrolytic dissociation of the polar groups: NP, in a polar medium; P_{M^+}, in an alkali polar medium; $P_{Ba^{2+}}$, in an alkali earth polar medium.

strong mutual repulsion between the divalent barium ion and the dissociated sodium ion associated with the polar barium structural unit $[Na^+B^-O_{4/2}Ba^{2+}]$ also facilitates this increase in Ψ_Φ. The repulsion leads to the increase in the potential energy of dissociation of the sodium ion by

$$\Delta W_+ = 330 \left(\frac{z}{d + r_{Ba^{2+}} - r_{Na^+}} - \frac{1}{d} \right) \text{kcal/mole},$$

where d, and the radii of the ions, are in Å [7]. This also occurs in the presence of other alkali earth ions [8].

We have to remember also that alkali earth ions considerably polarize the surrounding alkali dipolar structural units and thus sharply decrease the autosolvation effect of the alkali cations. This makes the electrolytic dissociation of alkali cations difficult even when they are present in a considerable concentration in the glass.

5. The expression considered here for the degree of ionization is to be preferred to the molar percentage expression. When the composition of a glass is expressed in molar percentages using conventional molecular weights of the compounds B_2O_3 and SiO_2, the equivalence of B_2O_3 to two $BO_{3/2}$ structural units is not taken into consideration. The glasses, 1) 25% PbO + 75% SiO_2 and 2) 25% PbO + 75% B_2O_3 which appear similar if expressed in molar percentages, in fact correspond to fundamentally different physical chemical states, distinct in the degree of ionization of the covalent vitreous network; $\alpha_1' = 0.66$ and $\alpha_2' = 0.33$. Similarly, the glasses expressed as: 1) 32% Na_2O + 68% SiO_2 and 2) 32% Na_2O + 68% B_2O_3 correspond respectively to $\alpha_1' = 0.94$ and $\alpha_2' = 0.47$.

With an increase in the valency of the cation we should expect an increase in the Ψ_Φ energy roughly proportional to the electrovalence z of the cation [7].

In fact, in barium and zinc borate glasses rich in metal oxide the parameter is about 70-80 kcal/mole instead of 35-40 kcal/mole as in borate glasses with a high content of univalent alkali ions [4].

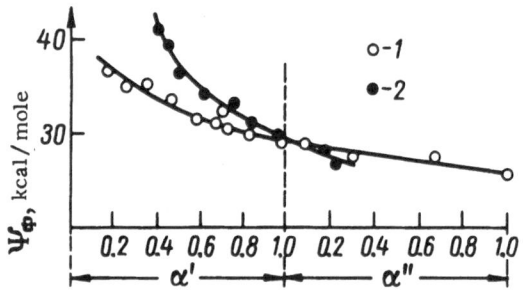

Fig. 7. The dependence of Ψ_Φ on the multiplicity of ionization α in silicate glasses (stable state, 200°C): 1) Na; 2) K.

Shul'man's experimental data in [20] for the conductivity of trivalent aluminum oxide (alundum) yields the temperature dependence

$$T < 1400° K, \quad \varkappa = 10^{-6.2} \exp\left(-\frac{4750}{T}\right), \quad (8a)$$

$$T > 1400° K, \quad \varkappa = 10^{1.5} \exp\left(-\frac{30000}{T}\right). \quad (8b)$$

According to Frenkel' in [21], the pre-exponential factor will be 10^3 when the concentration of trivalent aluminum ions is $n_0 = 5 \cdot 10^{22}$ ions/ml. The temperature dependence at high temperatures in Eq. (8) agrees with this value. We may, therefore, consider the dissociation energy as $\Psi_\Phi = 2AR = 119$ kcal/mole, where $A = 30,000$. This provides a basis for the calculation of the high temperature conductivity of alumina, which is due to dissociated trivalent aluminum cations with a dissociation energy three times that of univalent cations.

The low temperature conductivity (Eq. 8a) is probably due to the very small sodium impurity content encountered everywhere. According to Frenkel' the pre-exponential factor corresponds to the impurity sodium ion concentration, $n_0 = 3 \cdot 10^{20}$ C = 1014.3 ions/ml, 3 · 10^{-10} mole/ml (Na ion content of crystalline NaCl is $4 \cdot 10^{-2}$ mole/ml, that of Al ions in Al_2O_3, $7 \cdot 10^{-2}$ mole/ml). The sodium impurity is 1 in 10^8 of the base material. The dissociation energy of the sodium impurity is small, $\Psi_\Phi = 2.4750R = 18.8$ kcal/mole.

6. Proceeding from the previous position we can now consider borosilicates with a multiionized structure.

Figure 7 shows the dissociation energy of sodium and potassium silicates as a function of the extent of ionization of their structure (the experimental data are from Kuznetsov and Mel'nikova [22] and Seddon, Tippett and Turner [23]). The declining rate of decrease of energy of dissociation with the increase in the polarity of the medium is in agreement with the general principles discussed above.

Our experimental results for Ψ_Φ and P_e for divalent Ba and Zn are shown in Fig. 8 [18].

For the strongly ionized lead silicates the data on density [24] and the conductivity measurements of Evstrop'ev et al. [25] make it possible to calculate P_e and the dissociation energy.

The results of these calculations are shown in Table 3.

The theoretical value of $P_\sigma = 3.83 + 2 \log z = 4.4$, may be as low as 3.4 if the thermal vibrational frequency of the lead ions is reduced to $\nu \simeq 10^{12}$ sec^{-1} [6]. As can be seen from Fig. 8, P_e reaches values beyond the theoretical limit, having monotonically increased with the rise in the degree of ionization of the structure by more than three orders of magnitude. An increase in the average displacement distance ("mean free path") of the lead ion with a decrease in the activation energy must be excluded. The constancy, or even slight increase in the dissociation energy Ψ_Φ, excludes this possibility. We have here the rare case of a very weak concentration dependence of the dissociation energy while the statistical factor C increases markedly with concentration. It is not impossible that this is due to the inclusion of increasingly polarizable lead atoms in a covalent vitreous network [26]. The dissociation energy does not change as a result of the absence of ionization of the structures. The increase in the vacancy sites for dissociated cations around the valence-bonded lead atoms produce the increase in the statistical factor.

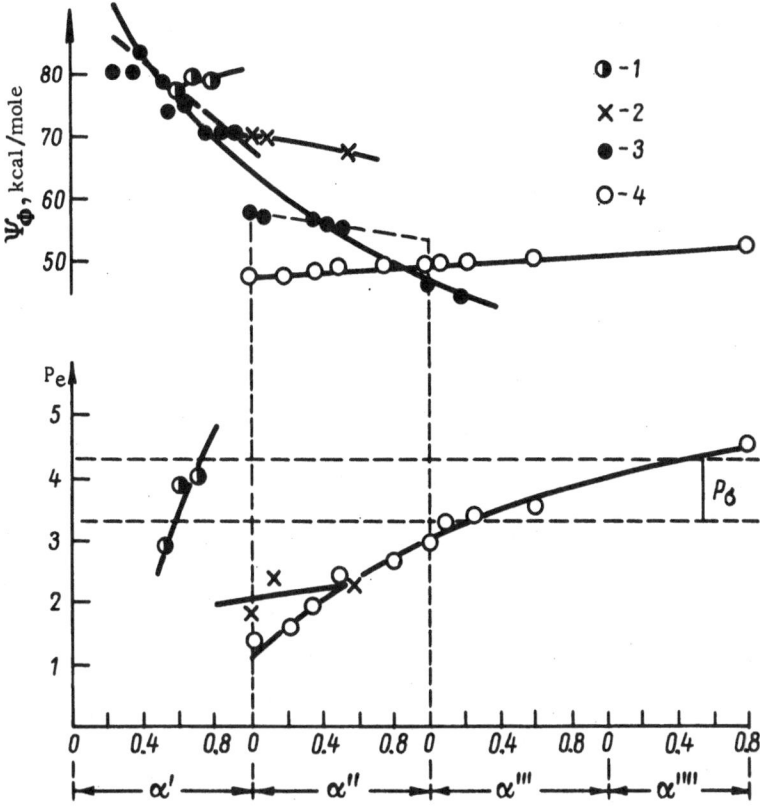

Fig. 8. The dependence of Ψ_Φ and P_e on the multiplicity of
ionization α in borate glass: 1) Ba; 2) Zn; 3) Pb; 4) lead silicate glasses.

According to the data given by Evstrop'ev and coworkers [25] and shown in Table 4, no
such anomalies are observed in lead borates. Due to the absence of density data for lead borates the calculations in Table 4 are incomplete. The theoretical value of $P_\sigma = 3.64 + \log z = 4.2$
may be as low as 3.2 if the oscillation frequency is lower. Assuming a density of about 6 it is
possible for the lead ion concentrations in glasses Nos. 1-16 [25] to be within the limits $0.012 \le$
$[Pb] \le 0.04$ mole/ml. Hence we obtain an experimental value $2.8 \le P_e = 0.4343B - \log[Pb] \le 3.5$
in satisfactory agreement with the approximate theoretically calculated limits of P_σ.

The dissociation energy, Ψ_Φ may be considered to decrease slowly and monotonically as
the degree of ionization of the system increases. However, we cannot exclude the possibility
that the decrease occurs stepwise, with steps at the points $\alpha' = 1.00$ and $\alpha'' = 1.00$ when the elements of lower ionization disappear and units of higher ionization appear. The step in conductivity when $\alpha' = 1.00$ established in [25] argues in favor of a step in the dissociation energy Ψ
[27].

The temperature coefficient of conductivity gives an energy of 70-80 kcal/mole when
$\alpha' \le 1.0$ and only 50-60 kcal/mole when $\alpha'' \le 1.0$.

7. The rapid increase in the conductivity of borate and silicate glasses with an increase
in the alkali content is not due to a destruction of their network structure, as is sometimes suggested [28]. In the case of borate glass the addition of alkali oxides not only does not lead to a
reduction in the number of valence bonds; on the contrary exactly the reverse occurs. The number of bonds increases [16]. The melting temperature increases; the critical temperature increases; and according to our results borate glasses became more chemically stable (Fig. 9).

TABLE 3. Lead Silicate Glasses [25]

Glass No.	Mole %		δ, g/cm^3	$100[Pb]$, mole/ml	$100[Si]$, mole/ml	Degree of ionization			B	P_e	A	Ψ_Φ, kcal/mole
	PbO	SiO$_2$				α''	α'''	α''''				
12	33.8	66.2	4.6	1.3	2.65	0.02	—	—	−1.4	1.3	12100	48.0
11	37.1	62.9	4,8	1.5	2.51	0.18	—	—	−0.69	1.5	12000	47.5
10	40.2	59.8	5.2	1.7	2.50	0.34	—	—	0.23	1.9	12100	48.0
9	43.1	56.9	5.5	1.8	2.45	0.52	—	—	1.6	2.5	12400	49.4
8	47.3	52.7	5.7	2.0	2.20	0.80	—	—	2.1	2.6	12400	49.4
7	50.2	49.8	5.9	2.1	2.09	—	0.02	—	2.8	2.9	12400	49.4
1	51.4	48.6	6.1	2.2	2.07	—	0.12	—	3.5	3.2	12400	49.4
6	53.2	46.8	6.3	2.3	2.01	—	0.27	—	3.9	3.3	12400	49.4
5	57.1	42.9	6.5	2.4	1.84	—	0.66	—	4.9	3.5	12400	49.4
4	60.0	40.0	6.7	2.5	1.70	—	1.00	—	—	—	—	—
3	63.2	36.8	7.0	2.7	1.66	—	—	0.43	—	—	—	—
2	66.6	33.4	7.1	2.8	1.40	—	—	1.00	7.2	4.6	12900	51.2

TABLE 4. Lead Borate Glasses [25]

Glass No.	Mole %		Degree of ionization			B	A	Ψ_Φ, kcal/mole
	PbO	B$_2$O$_3$	α'	α''	α'''			
1	21.4	78.6	0.27	—	—	} 3.7	20200	80.5
2	21.5	78.5	0.27	—	—			
3	24.1	75.9	0.32	—	—	4.2	20200	80.5
4	28.0	72.0	0.39	—	—	} 5.5	20600	82.4
5	30.8	69.2	0.45	—	—			
6	33.3	66.7	0.50	—	—	5.3	20200	80.5
7	36.4	63.6	0.56	—	—	3.7	18800	75.0
8	39.5	60.5	0.65	—	—	4.4	18800	75.0
9	42.4	57.6	0.74	—	—	3.9	17900	71.5
10—1	44.0	56.0	0.79	—	—	4.6	17900	71.5
11—2	46.8	53.2	0.88	—	—	5.3	17900	71.5
12—6	51.0	49.0	—	0.04	—	2.3	14700	58.6
13–3	53.0	47.0	—	0.13	—	} 2.6	14370	57.2
14–3	56.0	44.0	—	0.27	—			
15–2	56.4	43.6	—	0.29	—			
16—1	61.0	39.0	—	0.56	—	2.7	13800	55.0
17—2	67.2	32.8	—	—	0.05	0.7	11950	47.6
18—2	69.0	31.0	—	—	0.23	0.8	11600	46.2

At the same time as the conductivity of the borate glasses is increasing by 10^5 with an increase in Na content, the solution rate is decreasing by 10^2 [9, 13, 29]. In this case the viscosity and conductivity are independent of each other [11, 30].

8. An important phenomenon occurring in vitreous materials in the melt is often ignored [26, 28, 31]. We refer here to an association of polar structural units in a vitreous anhydride

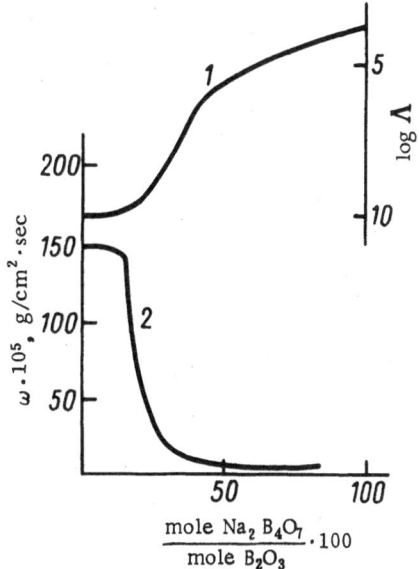

Fig. 9. The dependence of log molar conductivity, log Λ (1), and solution rate (2) of sodium borate glasses, on the molar ratio $Na_2B_4O_7/B_2O_3$.

medium with a low dielectric constant. In low dielectric constant media with low electrolyte concentration a minimum in molar conductivity is observed, as a result evidently of the association of polar particles [32].

It is easy to see that the association of two dipoles $B^-O_{4/2}Li^+$ or $B^-O_{4/2}K^+$ is accompanied by a significant decrease in their potential energy given by

$$-\frac{28.9\,p^2}{\varepsilon r^3} \leqslant \Delta W \leqslant -\frac{14.45\,p^2}{\varepsilon r^3} \text{ kcal/mole,}$$

where p is the dipole moment; ε is the dielectric constant; and r is the distance between dipoles, in Å. Taking into account the distance between the dipoles [7] and the values $3 \leq \varepsilon \leq 4$, and $r \simeq 3$ Å, we find that $8.4 \leq W \leq 39.8$ kcal/mole.

The association of the polar units increases when alkali-earth divalent metals are involved and may even produce opalescence [18, 33]. This type of association is probably present in opalescent sodium borosilicate glasses. The calculation made by the author from the data of Levin, Zhdanov, and Porai-Koshits [34], for the dependence of the intensity of the Rayleigh scattering on the time for which glass Na 7-23 was held at various temperatures, gave a value of 39.3 kcal/mole for the activation energy.

The minimum in the molar conductivity at $\gamma = 15$ [4], and the minimum in the high frequency dielectric loss noted by Stevels [35] in sodium and lithium borosilicate glasses when $5 \leq \gamma \leq 7$, are due to the association of dipolar structural units.

9. The decrease in the conductivity with a decrease in their alkali content cannot be the simple results of a "condensation" of the units. According to our data [4, 5] the packing density of the structural units in lithium borate glasses even increases with an increase in their alkali content. The rise in the volume concentration of boron atoms confirms this view (Fig. 10). In other borates the decrease in packing density of the borate structural units begins when $8 \geq \gamma > 6$, and in this case the decrease is enhanced by the greater size of the alkali ions. An analogous effect is observed in silver and barium borate glasses.

When boron changes from the trivalent to the quadrivalent state the distance between the centers of the boron and oxygen atoms increases from 1.36 to 1.53 Å [36]; i.e., by 12.5%. This would have had to be accompanied, when the stage of single ionization had been reached, by a decrease in the volume concentration of the boron atoms of 42%. In fact, the loose trigonal boric anhydride medium becomes densified by the addition of metal oxides due to both the increasing content of penetrating metal ions and oxygen atoms and to the increased coordination of boron atoms around each other. This explains the increase in the volume concentration of boron atoms in the lithium glasses, and the slow decrease in their concentration in other glasses. Simultaneously the structure is strengthened by the increasing density of the covalent bonds in the matrix.

Figure 11 shows the volume concentration of silicon and oxygen in silicate glasses with increasing ionization of the covalent units (experimental data [24, 37]; calculation [4]). In Fig. 11 the continuity of the sodium and lead silicate curves, in spite of the difference in the electrovalence of the ions, should be noted. This indicates the unique effect of the degree of ionization

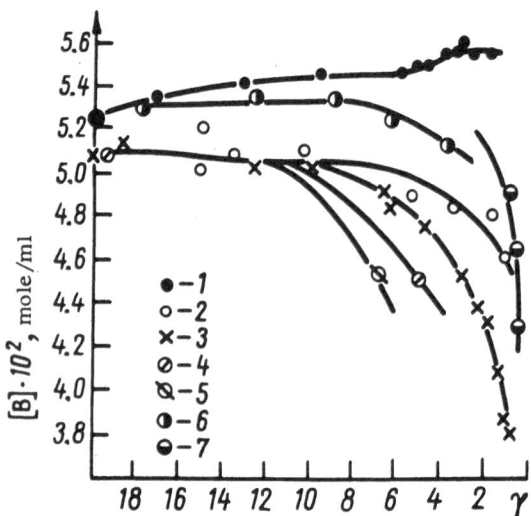

Fig. 10. The dependence of the molar volume concentration of boron atoms on γ in borate glasses: 1) Li; 2) Na; 3) K; 4) Rb; 5) Cs; 6) Ag; 7) Ba.

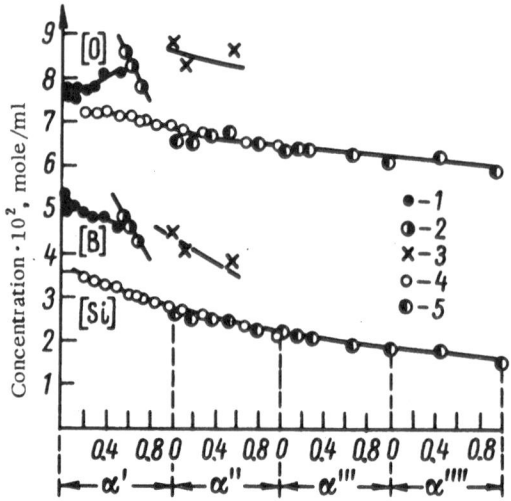

Fig. 11. Dependence of molar volume concentration of boron, silicon, and oxygen atoms on the degree of ionization of the structure α. In borate glass: 1) Na, 2) Ba, 3) Zn; in silicate glass: 4) Na, 5) Pb.

of the covalent units on their physico-chemical properties. At the same time the existence of this unique effect of the degree of ionization of the structure in sodium and lead silicate is in conflict with the anomalous concentration dependence of P_e and Ψ_Φ in lead silicates. Evidently the latter requires further investigation.

Conclusions

1. Following consideration of the structural chemical features, the dependence of the molar conductivity of borosilicates, $\Lambda = C \exp\left(-\dfrac{\Psi_\Phi}{2RT}\right)$, on the concentration of metallic ions has been investigated.

2. The concentration dependence of the molar conductivity has been shown to be determined, to a first approximation, by the concentration dependence of the dissociation energy Ψ_Φ. Lead silicate is exceptional and requires further study.

3. An explanation, based on the change in the dissociation energy Ψ_Φ, is given for the slow change in molar electrical conductivity in the low alkali ion concentration region, and the steep increase in the high concentration region.

4. An explanation has been proposed for the slower change with concentration of the dissociation energy Ψ_Φ in the polar medium of silicates compared with the more rapid change of this parameter in borates.

5. New mathematical expressions are presented for the concentration dependence of the dissociation energy for alkali ions in the polar borosilicate medium, and for the molar conductivity of borosilicates

$$\Lambda = C \exp\left[-\frac{700 + 0.25\beta\gamma}{T}\right].$$

6. The reasons are examined for the lowering of the alkali conductivity of glass in the presence of alkali earth oxides.

7. Taking borosilicates and alumina as examples, the dissociation energy and the electrovalency of the conducting cation are shown to be proportional to each other.

8. An increase in the conductivity is not caused by a loosening of the glass structure, since a decrease in conductivity is not a consequence of the 'condensation' of the structural units.

9. The change in the degree of ionization of the covalent glass structure and the association of polar ionic structural units during the melting of glasses, is the basis for the concentration dependence of electrical conductivity.

References

1. R. L. Myuller, Zh. Fiz. Khim., 6:616 (1935); R. L. Myuller (Müller), Phys. Z. Sowjetunion, 1:407 (1932), Acta Physicochim. URSS, 2:103 (1935); R. J. Maurer, J. Chem. Phys., 9:579 (1941); O. L. Anderson and D. A. Stuart, Ind. Eng. Chem., 46:154 (1954).

2. R. L. Myuller and B. J. Markin, Zh. Fiz. Khim., 5:1272 (1934); R. L. Myuller (Müller) and B. I. Markin, Acta Physicochim., URSS, 1:266 (1934).

3. S. A. Shchukarev and R. L. Myuller, Zh. Fiz. Khim., 1:625 (1930); S. A. Shchukarev (Schtschukarew) and R. L. Myuller (Müller), Z. Phys. Chem., A150:439 (1930).

4. R. L. Myuller, Zh. Tekhn. Fiz., 25:1868 (1955), ● this volume, p. 50.

5. R. L. Myuller, Zh. Tekhn. Fiz., 25:2428 (1955), ● this volume, p. 61.

6. R. L. Myuller, Zh. Tekhn. Fiz., 25:246 (1955), ● this volume, p. 24.

7. R. L. Myuller, Zh. Tekhn. Fiz., 25:1567 (1955), ● this volume, p. 43.

8. M. E. Spaight and J. D. Clark, J. Phys. Chem., 38:833 (1934).

9. R. L. Myuller (Müller), Phys. Z. Sowjetunion, 1:407 (1932).

10. S. Stegmaier and A. Dietzel, Glastechn. Ber., 18:304 (1940).

11. R. L. Myuller, Zh. Tekhn. Fiz., 25:236 (1955), ● this volume, p. 15.

12. R. L. Myuller, Zh. Fiz. Khim., 28:1193 (1954).

13. R. L. Myuller (Müller), Nature, 129:507 (1932).

14. R. L. Myuller, Uch. Zap. Leningr. Gos. Univ., No. 54, 159 (1940), ● this volume, p. 3

15. B. I. Markin, Zh. Tekhn. Fiz., 22:932 (1952).

16. R. L. Myuller, Zh. Fiz. Khim., 28:1954, 2170 (1954).

17. R. L. Myuller, Izv. Akad. Nauk SSSR, Ser. Fiz., 4:607 (1940), ● this volume, p. 170.

18. R. L. Myuller, Zh. Fiz. Khim., 6:616 (1935); R. L. Myuller (Müller), Acta Physicochim., URSS, 2:103 (1935); R. L. Myuller and B. I. Markin, Zh. Fiz. Khim., 7:592 (1936); R. L. Myuller (Müller) and B. I. Markin, Acta. Physicochim, URSS, 4:471 (1936).

19. A. F. Val'ter, M. A. Gladkikh, and K. I. Martyushov, Zh. Tekhn. Fiz., 10:1593 (1940).

20. G. I. Skanavi, Dielectric Physics, GITTL, Moscow—Leningrad (1949), p. 280; A. R. Shul'man, Zh. Tekhn. Fiz., 10:1173 (1940).

21. Ya. I. Frenkel', Z. Phys., 35:657 (1926).

22. A. Ya. Kuznetsov and I. G. Mel'nikova, Zh. Tekhn. Fiz., 24:1204 (1950).

23. E. Seddon, E. Tippett and W. S. Turner, J. Soc. Glass Technol., 16:459 (1932).

24. K. S. Evstrop'ev and N. A. Toropov, Silicon Chemistry and the Physical Chemistry of Silicates, Promstroiizdat, Moscow (1950), p. 325, Fig. 243.

25. K. S. Evstrop'ev, A. Ya Kuznetsov, and I. G. Mel'nikova, Zh. Tekhn. Fiz., 21:104 (1951); Zh. Fiz. Khim., 25:1318 (1951).

26. J. E. Stanworth, J. Soc. Glass Technol., 32:154 (1948).

27. R. L. Myuller, Zh. Fiz. Khim., 30, 1146 (1956).

28. G. I. Skanavi, Dielectric Physics, GITTL, Moscow–Leningrad, 1949; J. M. Stevels, Progress in the Theory of the Physical Properties of Glass, New York–Amsterdam–London–Brussels (1948); O. V. Mazurin, Candidate's Dissertation in Chemical Science, Leningr. Tekhnol. Inst. (1953).

29. R. L. Myuller and Ts. V. Vainshtein, Zh. Fiz. Khim., 7:364 (1936); R. L. Myuller (Müller) and Ts. V. Vainshtein (C. W. Wainstein), Acta Physicochim, URSS, 3:465 (1936).

30. R. L. Myuller, Zh. Prikl. Khim., 28:363, 1077 (1955).

31. A. A. Appen, Zh. Tekhn. Fiz., 23:1870 (1953).

32. R. Waldren, Molekulargrössen von Elektrolyten in nichtwässerigen Lösungsmitteln, Dresden–Leipzig (1923); L. Ebert, Handbuch der Experimentalphysik., Leipzig (1932); R. Wolf, Angew. Chem., 67:89 (1955); C. A. Krauss, J. Phys. Chem., 58:678, 683 (1954).

33. W. Guertler, Z. anorg. Chem., 40:337 (1904).

34. D. I. Levin, S. P. Zhdanov, and E. A. Porai-Koshits, Izv. Akad. Nauk SSSR, No. 1, p. 31 (1955).

35. J. M. Stevels, Philips Res. Repts., 7:161 (1952).

36. J. Biscoe and B. E. Warren, J. Amer. Ceram. Soc., 21:287 (1938).

37. C. J. Peddle, J. Soc. Glass Technol., 4:9 (1920).

IX

Degree of Dissociation and Cation Mobility
in Glasses with One Kind of Ion

1. Electrolytic dissociation in glasses is very low [1] and therefore the equilibrium constant can be expressed in terms of the concentration without using chemical activities

$$K = \exp\left(-\frac{\Delta\Phi}{RT}\right) = \frac{n_d^2}{n_0^2} = \alpha^2. \tag{1}$$

Here $\Delta\Phi$ is the dissociation energy; n_d the vacancy concentration ζ^-, equal to the concentration of the dissociated cations M^+, n_0 is the total concentration of alkali ions M^+, equal to the concentration of polar units $\zeta^- M^+$ where ζ^- is the negatively ionized structural units of the vitreous network and α is the degree of dissociation.

The absolute vacancy mobility is equal to

$$w_t = -\frac{\vec{n}}{n_d} = \frac{\delta^2 \nu e}{3kT} \exp\left(-\frac{E_a}{RT}\right) \text{cm}^3/\text{sec} \cdot \text{v}.^\dagger \tag{2}$$

Here E_a is the activation energy required to shift a cation from one vacancy to the neighboring vacancy, δ is the distance of the shift, and ν is the oscillation frequency of the ion. Our final theoretical expression differs in principal from that proposed by Stevels [2] in that we take account of the degree of dissociation. The product of the degree of dissociation α and the absolute mobility w_t can be obtained from Eqs. (1) and (2) in the form

$$(w\alpha)_\tau = \frac{\delta^2 \nu e}{3kT} \exp\left(\frac{\Delta\Phi + 2E_a}{2RT}\right) = w \ \exp\left(-\frac{\Psi_\Phi}{2RT}\right) \text{cm}^3/\text{sec} \cdot \text{v}. \tag{3}$$

The exponential factor depends essentially upon the value of the dissociation energy and therefore, to a first approximation, $\exp(-\Psi_\Phi/2RT)$ gives the degree of dissociation. The factor ω_{0t}, which changes very little with temperature, is exclusively associated with the ion mobility.

The mobility term ω_{0t} is independent of the chemical composition of the glass and for borosilicates with univalent cations is obtained from Eq. (3) and is, within an order of magnitude, equal to

---

*R. L. Myuller, Fiz. Tverd. Tela, 2(6):1333 (1960).

$\dagger$ Equation 2 is derived from Eq. (5) on p. 34. As noted in the footnote on p. 34, $\vec{n} = \dfrac{\Delta n \delta}{H}$ where Δn is the excess number of ions or vacancies displaced in the field direction per cm^3 per sec. The absolute vacancy mobility is obtained by multiplying the ratio of vacancies displaced in the field direction per cc to their total number n_d, by the average distance they move (in cm), and dividing by the field strength (in V/cm): $W_T = \Delta n/n_d \cdot \delta \cdot 1/H$. Consequently, $W_T = \vec{n}/n_d$. (Editor's note)

$$w_{0t} = \frac{\delta^2 \nu e}{3kT} = 5.6 \cdot 10^{-3} \ \text{cm}^2/\text{sec} \cdot \text{v} \tag{3a}$$

(after inserting an average value for borate and silicate glasses of $\delta = 2.7 \cdot 10^{-8}$ cm [1], and for an average temperature of 500°K).

Thus we may separate Eq.(3) into its factors: the mobility factor $\omega_{0t} \neq f$ ([M], T) and the factor expressing the degree of dissociation exp $(-\Psi_\Phi/2RT) = f$([M], T).

The value of Ψ_Φ depends upon the concentration of polar structural units and for the glasses which will be considered below, in the stable state Ψ_Φ is independent of temperature.

The experimental data confirms these assumptions.

2. The experimental value of $(\omega\alpha)_e$ is determined from the specific electrical conductivity

$$\varkappa = n_0 e w \alpha = [M] \ Fw\alpha = \exp\left(B - \frac{A}{T}\right), \tag{4}$$

where [M] $= n_0 / N_0$ is the concentration of univalent cations in mole/ml (n_0 is the number of ions/ml and N_0 is Avogadro's number), $F = N_0 e = 9.65 \cdot 10^4$ C/g - eq. It is obvious that

$$(w\alpha)_e = \frac{\varkappa}{[M] \ F} = \exp\left(B - \ln [M] \ F - \frac{A}{T}\right), \tag{5}$$

or, analogous to Eq. (3),

$$(w\alpha)_e = 10 \ P_e - 4.98 \exp\left(-\frac{\Psi_\Phi}{2RT}\right) = w_{0e} \exp\left(-\frac{\Psi_\Phi}{2RT}\right) \text{cm}^2/\text{sec} \cdot \text{v}, \tag{6}$$

where $P_e = 0.4343B - \log [M]$ and $\Psi_\Phi = 2AR$.

From Eqs. (3), (3a), and (6) in logarithmic form, we get

$$\log (w\alpha)_t = \log w_{0t} - \frac{\Psi_\Phi}{4.6RT} = -1.25 - \frac{\Psi_\Phi}{4.6RT},$$
$$\log (w\alpha)_e = \log w_{0e} - \frac{A}{2.37} = P_e - 4.98 - \frac{\Psi_\Phi}{4.6RT}, \tag{7}$$

where Ψ_Φ is expressed in calories.

The agreement between the theoretical and experimental values of $(\omega\alpha)$ presumes above all, that $\omega_{0t} = \omega_{0e}$, i.e., $P_e = 4.98 - 1.25 = 3.7$ to within $\pm$ 1.

3. Table 1 shows the values of P_e calculated from the experimental data for the electrical conductivity of borosilicate glasses. We include here the data obtained by co-workers in our laboratory [3] and the results of other authors [4, 5]. The volume concentrations in silicate glasses were obtained by interpolation from the density values obtained by Peddle, Dietzel, and Sheybany [6].

The calculated values of the basic parameters for silicate glasses are given in Table 2. Comparison of the experimental values for P_e in Table 1 with the theoretical value of 3.7 $\pm$ 1 calculated above, confirms the satisfactoriness of the theoretical calculation for the absolute ion mobility W_{0t} in glass. We deduce simultaneously from Eq. (3a) the independence of the mobility term on the chemical composition of the glass and the weak dependence of mobility on temperature. Table 1 also shows that the various authors have obtained different values for P_e. These differences are probably due to differences in the glass melt and in the cooling conditions. For lithium borate glasses [3] and sodium silicate glasses [4, 5] this difference in the data for P_e reaches 12%.

TABLE 1. Values of P_e , L, and L'. (Calculated from the Experimental
Data [3-5])

Glass	P_e	L	L'	References
Lithium borate	4.4 ± 0.3	5.7 ± 0.1	−	[3]
Lithium borate	3.9 ± 0.2	5.3 ± 0.1	−	[3]
Sodium borate	3.4 ± 0.3	4.9 ± 0.1	−	[3]
Potassium borate	3.5 ± 0.2	5.1 ± 0.2	−	[3]
Silver borate	4.5 ± 0.1	5.0 ± 0.1	−	[3]
Thallium borate	3.7 ± 0.1	5.0 ± 0.1	−	[3]
Sodium borosilicate	3.5 ± 0.1	5.1 ± 0.1	(14.0 ± 0.7)	[3]
Potassium borosilicate	3.5 ± 0.1	5.3 ± 0.1	(15 ± 1)	[3]
Sodium silicate	3.6 ± 0.1	−	11.8 ± 0.2	[4]
Sodium silicate	3.2 ± 0.1	−	10.7 ± 0.3	[5]
Potassium silicate	3.3 ± 0.2	−	12.1 ± 0.3	[5]

TABLE 2. Silicate Glasses (Calculated from the Experimental
Data [4-6])

Glass	$[M_2O]$, mole %	d, g/cm³	100 [M], mole/ml	100 [Si], mole/ml	B_{10}^*	P_e	Ψ_Φ, $\frac{kcal}{mole}$	L'
Sodium	27	2.45	2.12	2.95	1.52	3.2	28.8	11.0
	30	2.47	2.45	2.85	1.54	3.2	27.5	10.9
	33.3	2.49	2.73	2.74	1.52	3.1	25.7	10.5
	36	2.51	2.92	2.64	1.50	3.0	24.9	10.3
	40	2.53	3.30	2.50	1.57	3.1	24.0	10.2
	45	2.55	3.80	2.31	1.95	3.4	24.5	10.8
	48	2.57	4.00	2.19	1.68	3.1	22.5	10.1
Potassium	15	2.36	1.09	3.08	1.55	3.5	38.4	12.4
	20	2.41	1.43	2.89	1.18	3.0	34.0	11.8
	23	2.43	1.64	2.76	1.55	3.3	33.3	11.9
	27	2.45	1.89	2.61	1.22	2.9	34.2	12.7
	33.3	2.48	2.32	2.32	1.91	3.5	29.9	11.7
	40	2.51	2.72	2.04	2.00	3.6	29.5	12.0

* See p. 105.

The energy Ψ_Φ, which is closely related to the dissociation energy $\Delta\Phi$, depends upon the volume concentration of the polar units $[\zeta^-M^+]$ which is in practice equal to the concentration of metallic ions [M] in the glass [7]. This functional relation in borate glasses is in agreement with the relationship $\Psi_\Phi [M]^{1/2} = L$ [8] which we have proposed. In Table 1, we give the average values of $L \approx 5$ for borate and borosilicate glasses; these values were obtained from the data for conductivity determined by the co-workers in our laboratory [3].

For silicate glasses a slight modification is noted in the law $\Psi_\Phi [M]^{1/4} = L' \approx 12$ [10]. A more gradual decrease in the dissociation energy with an increase in the number of polar units in the glass is determined, apart from the indicated role of the oxygen atom concentration, by the more limited possibilities in the induced orientation of silicate dipoles with trigonally located potential wells. As we know, the potential wells of polar borate units have four possible tetrahedral locations.

In Table 1 we give the mean values of L' for silicate glasses. We have calculated them according to Table 2 using conductivities determined by various author's [4-6]. Different authors have found slightly varying values for P_e and similarly for L and L'. For lithium borate glasses there is a difference of 7% [3], and for sodium silicate glasses it reaches 10% [4, 5]. This corresponds to fluctuations in Ψ_Φ of about ±3 kcal/mole.

The specific difference between the values of L and L' permits us to establish the predominance of the dissociation of polar borate units in borosilicate glasses. In fact, according to the data of Table 1, in borosilicate glasses the regularity of L that is characteristic of borate glasses is preserved. This also points to the presence in these glasses of an accumulation of alkali borate polar units. The formation of microporous glasses is another related phenomenon.

4. The analysis above of the magnitude of the mobility term ω_{0t} and of Ψ_Φ which determines the degree of dissociation, holds for glasses in the stable state. These physical characteristics of glass are retained for glass in the labile state in the critical temperature region. The steeper slope, observed in the latter case, of the curve of the log of the conductivity against the reciprocal of the temperature is associated with the gradual decrease in the dissociation energy. The dissociation energy decreases as a result of the appearance of secondary solvation due to Drude–Nernst electrostriction. This effect is produced by the increase in the statistically averaged time of the valence oscillations of the atoms in the vitreous network accompanying the switching over of the covalent bonds. With this change in free energy of dissociation the entropic ion-dissociation effect occurs.

The transition from the labile to the liquid state of the glass is accompanied by the disappearance of the entropic effect on ion dissociation, due to the onset in all the atoms of omnidirectional random thermal ascillations. Thus, it is characteristic of the conductivity of liquid glass that Ψ_Φ and P_e be independent of temperature. This independence of temperature is analogous to that observed for glass in the stable state. According to Eqs. (3a) and (7) the value of P_e must be close to 3.3 in the liquid glass.

The experimental data of Bockris, Kitchener, et al. [10] gives us the opportunity to compare the values of Ψ_Φ and P_e in sodium and potassium silicate glasses in the molten and stable solid states. Such a comparison has been made in Table 3, with the inclusion of the data of Tables 1 and 2 (100 [M] $\simeq$ 2-3 mole/ml) [4, 5].

The experimental values of P_e given above for liquid glasses agree with the theoretical values. We can consider that the limitation of the entropic effect to the region of existence of the labile state of the glass has been established.

The constancy of the value of P_e is indicative of the constancy of the values of ν and δ which enter into the expression for ω_0 (3a). This testifies to the preservation in the melt of the same mechanism for the shifting of the cations in the vacancies that is present in the solid glass. Speaking of the identical nature of the pre-exponential term in the expression for the specific conductivity of glass in the solid and molten states Kobeko, and later Mazurin [11], were close to confirming the identical nature of the ion mobility mechanism. The difference between their concepts and ours consists of the fact that they, and many other authors, have underestimated the decisive role of the dissociation energy of the polar units and the solvation phenomenon which produces the significant decrease in the dissociation energy. The decrease in the value

TABLE 3.

Glass	State			
	Liquid [10]		Solid stable (compare Tables 1 and 2)	
	P_e	Ψ_Φ	P_e	Ψ_Φ
Sodium silicate	3.1−3.3	22−26	3.4	26−33
Potassium silicate	$\simeq$ 3.2	16−24	3.3	29−38

of Ψ_Φ during the transition from the solid to the liquid state (see Table 3) is the result of a secondary solvation realized in the labile state.

In the light of the above considerations, it is a mistake to assume that there are significant numbers of ruptures of the covalent bonds between atoms ("network-breaking effect") during the melting of the glass, and that an increase in the ion mobility is the result of the appearance of "free spaces" [10]. The assumption that there is a complete degree of dissociation of the polar units in molten silicates also appears to be erroneous [12]. The constant numerical value of the mobility factor ω_{0t} and of the associated parameter P_e indicates that the cation mobility mechanism remains unchanged in glasses both in the solid and molten states. The high value of Ψ_Φ points to the absence of significant dissociation of the polar structural units.

The remarks on mobility and ion dissociation in glasses which apply to the electrical conductivity apply in equal measure also to diffusion. In this connection, for diffusion one should make use of the expression

$$D = D_0 \exp\left(-\frac{\Psi_\Phi}{2RT}\right),$$

where the quantity Ψ_Φ which is closely related to the dissociation energy must replace the quantity $2E_a$, the activation energy of ion mobility, which has been introduced without foundation.

Summary

1. We have demonstrated the possibility of theoretically computing the ion mobility term, ω_{0t}, in glass containing one type of ion. The value of Ψ_{0t} is effectively independent of the chemical composition of the glass and changes very little with temperature.

2. The degree of ion dissociation in glass is determined by the parameter Ψ_Φ which is numerically close to the ion dissociation energy in glass. The value of Ψ_Φ decreases regularly with an increase in the number of polar units in the glass. The nature of this phenomenon is associated with the solvating action of the polar units on dissociated ions and vacancies.

3. The nature of the parameters P_e and Ψ_Φ is preserved during the transition of the glass from the stable solid to the liquid state.

References

1.   Ya. I. Frenkel', Z. Phys., 35:652 (1926); R. L. Myuller, Zh. Fiz. Khim., 6:616 (1935); Acta Physicochim. URSS, 2:103 (1935); Zh. Priklad. Khim., 28:363 and 1077 (1955); Zh. Tekhn. Fiz., 25:236, 246, 1556, 1868 (1955), 26:2614 (1956), ● this volume, pp. 15, 24, 33, 50, 79.

2.   J. M. Stevels, Progress in the Theory of the Physical Properties of Glass (N. Y.-Amsterdam—London—Brussels, 1948); Handbuch der Physik, 20:350 (1957).

3.   S. A. Shchukarev and R. L. Myuller, Zh. Fiz. Khim., 1:625 (1930); Z. phys. Chem., A150:439 (1930); R. L. Myuller and B. I. Markin, Zh. Fiz. Khim., 5:1262 (1934); 7:592 (1936); Acta Physicochim. URSS, 1:266 (1934); 4:471 (1936); B. I. Markin, Zh. Obshch. Khim., 11, 285 (1941); Zh. Tekhn. Fiz., 22:932 and 941 (1952); L. R. Takking and N. P. Shchegoleva, Uch. Zap. Leningr. Gos. Univ., 108:17 (1949); R. L. Myuller, Zh. Tekhn. Fiz., 25:1868 (1955), ● this volume, p. 50.

4.   E. Seddon, E. Tippett, and W. E. S. Turner, J. Soc. Glass. Technol., 16:459 (1932); R. L. Myuller, Zh. Tekhn. Fiz., 25:1868 (1955).

5.   O. V. Mazurin, Dissertation, Leningr. Tekhnol. Inst. (Leningrad. 1953); O. V. Mazurin and E. S. Borisovskii, Zh. Tekhn. Fiz., 27:275 (1957).

6.   C. J. Peddle, J. Soc. Glass Technology, 4:9 (1920); A. Dietzel and H.-A. Sheybany, Verres et Refractaires, 2:63 (1948).

7. R. L. Myuller, Zh. Tekhn. Fiz., 25:1567 (1955), ● this volume, p. 43.

8. R. L. Myuller, Phys. Z. Sowjetunion, 1:407 (1932); Nature, 129:507 (1932).

9. R. L. Myuller, Zh. Fiz. Khim., 6:616 (1935); Acta Physicochim. URSS, 2:103 (1935).

10. J. O'M. Bockris, J. A. Kitchener, S. Ignatowitcz, and J. Tomlinson, Trans. Farad. Soc., 48:75 (1952).

11. P. P. Kobeko, The Amorphous State, Izd. Akad. Nauk SSSR, Leningrad–Moscow (1952; O. V. Mazurin, Tr. Leningr. Tekhnol. Inst., 29:72 (1954).

12. O. A. Essin and P. V. Gel'd, The Structure of Glass, Izd. Akad. Nauk SSSR, Moscow–Leningrad (1955), p. 44.

X

Electrical Conductivity of Glasses Containing
Two Kinds of Alkali Ion*

1. It is well known that when half of the alkali oxide of one type in a glass is replaced by an alkali oxide of another type there is a significant decrease in the electrical conductivity and dielectric losses. A series of experimental studies have been devoted to this problem. Efforts have also been made to find a theoretical explanation for the observed minima in the dielectric loss and conductivity. However, there is not at present a generally accepted explanation of this phenomenon. Incidentally, one of the reasons for the lack of clarity with respect to this process is that frequently the quantitative analysis of experimental data is not sufficiently rigorous.

Gehlhoff and Thomas [1] were the first to note the minimum in the conductivity of glass which occurs on the gradual replacement of sodium oxide by potassium oxide in silicate glass. They located the minimum by an indirect method involving a determination of the temperature at which a preselected standard conductivity was observed. The basic defect of their method consisted in the use of the erroneous method of replacing the sodium oxide by potassium oxide on a weight percentage basis.

Specific conductivity relates to the conductivity of one cm^3 of the substance and is therefor determined by the number of ions in one milliliter and, in particular

$$[Na] = \frac{2d_1 P_{Na_2O}}{100 \cdot 62.0} \text{ and } [K] = \frac{2d_2 P_{K_2O}}{100 \cdot 94.2},$$

where d is the density of the glass and P_{Na_2O} and P_{K_2O} represent the percentage by weight of the respective oxides. When an alkali oxide is replaced, the volume concentration (in mole/ml) of alkali ions should remain unchanged:

$$[M] = [Na] + [K] = \frac{2d_{1,2}}{100} \left(\frac{P_{N_2O}}{62.0} + \frac{P_{K_2O}}{94.2} \right) = \text{const}$$

or there must be a strict allowance for the change in the total. It is clear from this equation that the replacement of one weight unit of sodium oxide requires the introduction of one and one-half times this weight of potassium oxide. In an analogous manner, during the replacement of lithium oxide it is necessary to introduce three times the weight of potassium oxide. If this is not so the conductivity will change not only as a result of the change in the relative sodium and potassium content of the glass but also as a result of the large change in the total amount of alkali ions. Gehlhoff and Thomas did not take this into account. It might easily happen that we would observe a decrease in the conductivity and dielectric losses, not only during the partial replacement of the sodium oxide by potassium oxide, but also during its complete replacement. This was observed by Bogaroditski and Fridberg [2] who found, instead of a minimum, the greatest decrease in the dielectric losses when there was complete replacement of the sodium oxide by potassium oxide in weight percent. Analogous use of the weight percent concentration can be noted in some of the latest works [3].

*R. L. Myuller, Fiz. Tverd. Tela, 2(6):1339 (1960).

TABLE 1

Glass No.	M_2O, mole %	B_2O_3, mole %	100 [B], mole/cm^3	100 [K], mole/cm^3	100 [Li], mole/cm^3
52	15	85	4.74	0.84	—
87	28	72	4.85	0.87	1.00
55	27	73	4.30	1.52	—
115	22	78	5.50	—	1.55

TABLE 2

Glass No.	Na_2O, wt.%	BaO, wt.%	B_2O_3, wt.%	Na_2O, mole %	1000 [Na], mole/ml	1000 [Ba], mole/ml	$-\log x$ (300 °C)
4	2.78	—	97.2	3.1	1.6	—	10.71
5	5.27	—	94.7	5.9	3.1	—	10.90
6	5.8	—	94.2	6.5	3.5	—	10.76
7	7.3	—	92.7	8.1	4.5	—	10.63
8	12.6	—	87.4	13.9	7.9	—	9.14
9	16.9	—	83.1	18.5	11.0	—	6.90
74	3.28	40.27	56.45	4.7	3.3	8.1	13.3
75	5.20	40.33	54.47	7.5	5.2	8.1	12 0
76	10.45	39.8	49.75	14.7	10.8	8.3	9.3
77	12.8	40.48	46.72	18.0	13.2	8.4	8.1

Often, during the replacement of the alkali oxides, the investigators keep constant the molar percentage concentration of the total alkali oxides. It is not difficult to notice that in the case, for example, of sodium potassium silicate or borate glasses this is equivalent to the condition ([Na] + [K])/[Si] = const or ([Na] + [K])/[B] = const.

Keeping the molar percentage of the sum of alkali oxides constant does not maintain a constant number of alkali ions per unit volume of glass. The volume concentration of ions in the glass may, in this case, change significantly with a change in the volume concentration of boron or silicon atoms.

According to our data [4], when lithium oxide is added to potassium borate glass (No. 52 in Table 1) and we obtain lithium potassium glass (No. 87), the molar percentage of boric anhydride decreases by 13% but the atomic volume concentration of the boron increases 2%. The molar percentage of alkali oxides increases 1.87 times and atomic volume concentration by 2.23 times. For a complete replacement of the potassium by lithium (glasses Nos. 55 and 115 in Table 1), the molar percentage of boric anhydride increases by 5% and accordingly the molar percentage of alkali oxide decreases by 5% but the volume concentration of alkali ions increases by 1.5% as the result of the 28% increase in the volume concentration of borate structural units.

Ignoring the changes in the packing density of the borate and silicate units in the glass by using the molar percentage relationships, leads to serious misunderstanding. Figs. 1-3 serve as an illustration. They present the data obtained in our laboratory [5] for the conductivity of sodium and sodium barium borate glasses after the recalculation of the concentrations according to Table 2.

A true comparison of the data regarding the conductivity of sodium borate glass before and after the introduction into the glass of barium oxide is given in Fig. 1 in which the specific conductivity is plotted with respect to the volume concentration of sodium ions. As we see, upon the introduction of about 40% by weight of barium oxide (100 [Ba] = 0.82 ± 0.01) into the sodium borate glass, the electrical conductivity of the latter for 1000 [Na] = 10, decreases 100 times. With an incorrect choice of variable in the form of the molar percentage of Na_2O (Fig.2)

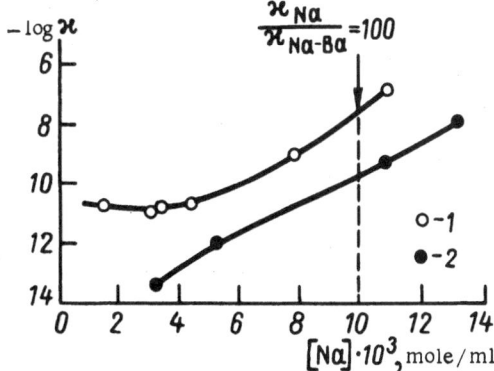

Fig. 1. The electrical conductivity of sodium borate glasses as a function of the volume concentration of sodium ions: 1) in the absence of barium oxide; 2) in the presence of barium oxide.

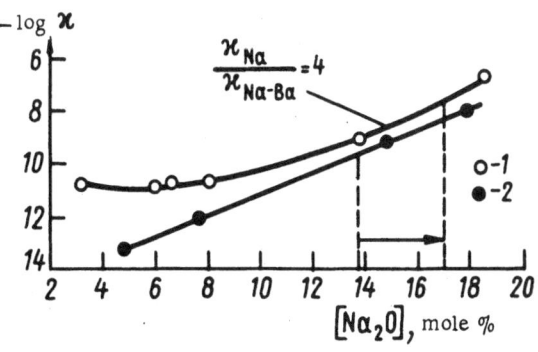

Fig. 2. The electrical conductivity of sodium borate glasses as a function of the molar percentage of sodium oxide: 1) in the absence of barium oxide; 2) in the presence of barium oxide.

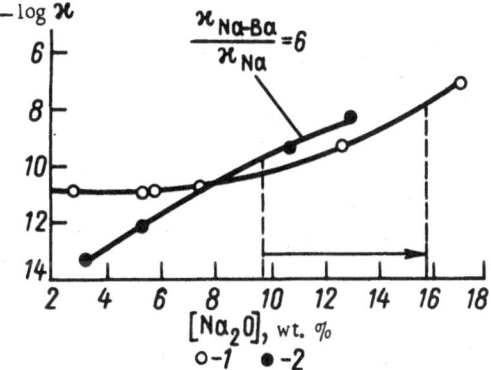

Fig. 3. The electrical conductivity of sodium borate glasses as a function of the weight percentage of sodium oxide: 1) in the absence of barium oxide; 2) in the presence of barium oxide.

we would get an erroneous judgement regarding the lowering of the sodium conductivity after the introduction of barium oxide which would be fourfold. This is explained by the decrease from 17.0 to 13.6 in the molar percentage of sodium oxide in the glass on the introduction of barium oxide whilst maintaining the same volume concentration 1000 [Na] = 10 mole/ml. A completely distorted picture is obtained if we make use of the weight % (Fig. 3). We get an absurd intersection of the curves, caused by a 6% decrease in the weight % of the sodium oxide in the glass for the same volume concentration following the introduction of 40% barium oxide.

The considerations expressed here are sufficiently convincing to consider that we have established the necessity of using the volume concentration of the current carriers in the analysis of data for the specific electrical conductivity of complex glasses.

2. The minimum electrical conductivity of alkali borate glasses was established in our laboratory by Markin. He showed that for complex glasses the minimum corresponds to the conductivity minimum obtained by the superposition of the partial conductivities of the corresponding simple glasses [6]. We then found the physical explanation of such a "minimum-effect" [7]. The superposition of partial electrical conductivities indicates that there exists differentiated groupings of similar polar structural units in these glasses.

Simultaneously with our studies Skanavi and Martyushov and later Val'ter, and Gladkikh and Martyushov, and Lengyel [8], making use of the less felicitous molar percentage expression for the composition of the glasses, also established the presence of a minimum in the dielectric losses and conductivity of complex borate and silicate glasses. The minimum effect was called by Skanavi and Martyushov the "neutralizational effect." The latter designation is

TABLE 3

Glass No.	100 [M]	100 [B]	100 [O]	[A]	$\varkappa, \Omega^{-1} \cdot cm^{-1}$ (250 °C)
117	1.92	3.96	6.79	0.127	$7.6 \cdot 10^{-7}$
123	1.62	4.82	8.06	0.146	$8.5 \cdot 10^{-10}$
126	1.67	5.29	8.75	0.157	$2.5 \cdot 10^{-8}$

unfortunate from a physicochemical point of view. The nature of the "minimum-effect" has nothing in common with the neutralization of two alkali ions. The ions of the alkali metals are so close to each other in their physicochemical properties that it is impossible to attribute mutual neutralizing properties to them.*

There is also a tendency on the part of some investigators to attribute the minimum-effect to a decrease in the glass of the percentage of "voids" according to Smekal, or "a priori vacancies" (Leerstellen, Fehlerstellen), or to obstructions which are supposed to arise during the translational motion in the glass of the various types of cations. We know, since the publication of Warburg's paper [9], that during electrolysis large radius cations very poorly occupy the "vacancies" at the anode which are left by smaller cations. With this as a basis Lengyel [10] computed the probability that large cations would get stuck in the "vacancies" left by the smaller cations. The calculations showed that in this case there must be a rapid cessation of all conductivity, contrary to the experimental results for electrical conductivity. The weakness in Lengyel's calculation consists of the fact that he ignored the random thermal motion of the "vacancies" which prevents the cations from sticking and opens circuitous paths for them. Also he did not take into account the differences between the injection of foreign cations into the glass at the anode accompanied by changes in the composition of the glass and the omnidirectional thermal motion of the vacancies and cations which were not accompanied by changes in the composition of the glass.

In order to bring about agreement between his interpretation and the experimental results, Lengyel introduced the assumption that a stationary electrical conductivity was established when the blockage was overcome by "activation of the de-activated places" ("Actiwiert sich die desaktiwierte Stelle"). Lengyel's mathematical expression for the conductivity of glass includes three empirical constants. By trial and error selection of these constants one can, with difficulty, approximate the experimentally obtained value of the conductivity.

It is clear from our explanation that Lengyel's hypothesis regarding the decisive role of ion blockage in the minimum effect has been theoretically contradicted and has proved to have no experimental foundation. Kobeko and Stevels put forward the theory that an increase in the structural density of the glass was the reason for the minimum effect [11]. There is no such connection in reality. For example, according to the data obtained in our laboratory [12] for lithium potassium borate glasses there is a minimum at which the conductivity has decreased 30–900 times while at the same time there is a monotonic increase in the packing density [A], in the boron atoms [B], oxygen atoms [O], and alkali metals [M] in g-atom/ml (see Table 3).

Comparing our data for lithium borate glasses [4], we note an increase of 10^6 in the specific electrical conductivity for a 27% increase in the total number of boron, oxygen, and lithium atoms per cm^3.

The inconsistency in the theory involving spatial mechanical interference during conductivity processes in glass is obvious from the data presented in the preceding paper [13]. Analysis

---

* At a later date Myuller suggested the term "poly-alkali effect" for this phenomenon. (Editor's note)

of the cation mobility term indicates that there is no necessity to introduce steric correction factors. The independence of the mobility term on the chemical composition of the glass shows the constancy in the value of the cation displacements and of their vibrational frequencies, and the small probability of a noticeable change in the activation energy of displacement of the dissociated ions.

3. For a sufficiently low total concentration of two alkali ions $[M]_s = [M'] + [M''] < 8 \cdot 10^{-3}$ mole/ml the associated polar groupings are small and are isolated by nonpolar boric anhydride. The electrical conductivity is limited by the transport of the dissociated cations in the nonpolar boric anhydride medium. The change in the energy Ψ_ϕ with concentration is insignificant and within the limits of 60-70 kcal [4]. As a result, the weak changes in the degree of dissociation with changes in the composition of the glass means there is no minimum effect in glasses with low alkali oxide content [6, 8]. The addition of alkali oxide to such glasses is accompanied by an increase in the conductivity and dielectric losses [8].

We note that in glasses with an increased alkali ion content $[M]_s > 8 \cdot 10^{-3}$ mole/ml there is a significant concentration dependence of the energy, with a maximum when there is an equivalent concentration of ions of the two types.

The minimum-effect in the electrical conductivity and the maximum in Ψ_ϕ in two-cation glasses is explained by the concentration dependence of simple glasses. The dependence of $\Psi_{\phi'}$ in two-cation glasses upon the composition is not explained by the concentration dependence of Ψ_ϕ in simple glasses in the absence of the assumption of a differentiation of the unlike polar structural units. For a similar but homogeneous mixture of the polar structural units the solvation and the energy Ψ_ϕ would be functions of the total concentration of both types of ions. Actually, in two-cation glasses, the curve of Ψ_ϕ has two branches each of which corresponds to the partial content of the prevailing alkali ion. Comparison of the corresponding values of Ψ_ϕ for single-cation and double-cation glasses confirms the differentiated association of the polar structural units in glass.

Summary

1. A critical analysis of the effect of ions on the electrical conductivity of complex glasses dictates the necessity of relating the conducting ion content to unit volume. The method sometimes used in such cases of expressing the composition of the glass in weight percentage units or in terms of molar percentage leads to erroneous electrochemical conclusions.

2. The explanation of the minimum effect in two-cation glasses in terms of spatial-mechanical retardation of the movement of the cations is erroneous.

3. The minimum-effect in the electrical conductivity of two-cation glasses is explained by the association of polar units which are differentiated according to the type of ion.

References

1. G. Gehloff and M. Thomas, Z. techn. Phys., 6:544 (1925).
2. N. P. Bogoroditskii and I. D. Fridberg, Zh. Tekhn. Fiz., 7:1905 (1937).
3. G. I. Skanavi and A. I. Demeshina, Zh. Tekhn. Fiz., 28:748 (1958).
4. R. L. Myuller, Zh. Tekh. Fiz., 25:1868 and 2428 (1955), ● this volume, pp. 50, 61; 26:2614 (1956) [Sov. Phys.-Tech. Phys., 1:2529] ● this volume, p. 79; B.I. Markin, Zh. Tekhn. Fiz., 10:66 (1940).
5. S. A. Shchukarev and R. L. Myuller, Zh. Fiz. Khim., 1:625 (1930); Z. Phys. Chem., A150: 439 (1930) R. L. Myuller and B. I. Markin, Zh. Fiz. Khim., 7:592 (1936); Acta Physicochim. URSS, 4:471 (1936).
6. B. I. Markin, The Electrical Conductivity of Complex Alkali Borate Glasses (Candidate's Dissertation), Leningr. Gos. Univ. (1938); Byull. Leningr. Otd. Vses. Khim. Obshchestva

im. D. I. Mendeleeva (May 9, 1938), Bulletin No.3-4, p. 40 (1939); Zh. Tekhn. Fiz., 10:66 (1940).

7. R. L. Myuller, Byull. Leningr. Otd. Vses. Khim. Obshchestva im. D. I. Mendeleeva, No. 6, p. 12 (1939), ● this volume, p. 169; Izv. Akad. Nauk ser. fiz, 4:607 (1940), ● this volume, p. 170

8. G. I. Skanavi and K. I. Martyushov, Zh. Tekhn. Fiz., 9:1024 (1939); A. F. Val'ter, M. A. Gladkikh, and K. I. Martyushov, Zh. Tekhn. Fiz., 10:1593 (1940); B. V. Lengyel, Glastechn. Ber., 18:177 (1940).

9. E. Warburg and F. Tegetmeier, Ann. Phys. (Wied.), 32:447 (1887); 35:445 (1888); 41:18 (1890); A. Günterschulze, Ann. Phys. (4), 37:435, 442, and 438 (1912); A. Heydweiller, F. Kopferman, Ann. Phys., 32(4):739 (1910); F. Quittner, Ann. Phys., 85(4):785 (1928); C. A. Krauss and E. H. Darby, J. Amer. Chem. Soc., 44:2783 (1922).

10. B. V. Lengyel and Z. Boksay, Z. phys. Chem., A203:93 (1954); A205:157 (1955).

11. P. P. Kobeko, Amorphous Substances, Izd. Akad. Nauk SSSR, Leningrad, 1952; J. M. Stevels, Philips Techn. Rundschau, 13:350 (1952).

12. N. I. Brodskaya and V. S. Tatarinova, Uch. Zap. Leningr. Gos. Univ. 54:241 (1940); L. R. Takking and N. P. Shchegoleva, Uch. Zap. Leningr. Gos. Univ., 108:17 (1949).

13. R. L. Myuller, Fiz. Tverd. Tela, 2(6):1333 (1960), ● this volume, p. 93.

XI

Degree of Dissociation and Cation Mobility in Glasses Containing Two Kinds of Ion

1. It is known that a minimum-effect exists in the electrical conductivity of borate and silicate glasses containing two types of cations with a total concentration $[M]_s \geq 8 \cdot 10^{-3}$ mole/ml. A complete classification of the physical character of this effect requires a careful, quantitative analysis of the basic parameters $P_e = B_{10} - \log [M]$ and $\Psi_\Phi = 4.6 A_{10}R$, where A_{10} and B_{10} are the coefficients in the conductivity equation $\ln \varkappa = -A/T + B$ after it has been converted to logarithm base 10.

The specific conductivity of a glass containing the single electrovalent cations M' and M'' is expressed in the form [1]

$$\varkappa = F[M'] (w\alpha)'_e + F[M''] (w\alpha)''_e. \tag{1}$$

Here $F = 9.65 \cdot 10^4$ C/g-eq; [M'] and [M''] represent the volume concentrations of the ions in the glass in mole/ml and $(\omega\alpha)'_e$ and $(\omega\alpha)''_e$ the experimental values of the products of the absolute mobilities ω and degrees of dissociation α for the corresponding cations. Separating out the exponential term, we can rewrite $(\omega\alpha)'_e$ and $(\omega\alpha)''_e$ in the form

$$\left. \begin{aligned} (w\alpha)'_e &= w'_{0e} \exp\left(-\frac{\Psi'_\Phi}{2RT}\right), \\ (w\alpha)''_e &= w''_{0e} \exp\left(-\frac{\Psi''_\Phi}{2RT}\right). \end{aligned} \right\} \tag{2}$$

Here each of the values Ψ'_Φ and Ψ''_Φ represents the sum of two magnitudes: a) the high dissociation energy $\Delta\Phi$ of a polar unit associated with the creation of a vacancy ζ^- and the emission of a cation M^+ into the space between the units, b) twice the small activation energy $2E_\alpha$ associated with the displacement of the cation from the potential well of the undissociated polar unit ζ^-M into the potential well of the neighboring dissociated vacancy ζ^- (which is adequate for the displacement of the vacancy). Analysis of the electrical conductivity of single-cation glasses has established that for $\gamma \leq 6$ † the value of the mobility term ω_{0e} is practically independent of the composition and corresponds to the theoretically computed value $\omega_{0t} = 5.6 \cdot 10^{-2}$ cm²/v · sec. As we will show later, in two-cation glasses there is no steric hindrance to the displacement of the cations to adjacent vacancies and to a first approximation the value of unit displacement, δ, is preserved. We can then equate $\omega'_{0e} = \omega''_e = \omega_e$ (for single-cation glasses) $\approx \omega_{0t} \neq f$ (composition). Experience with the replacement of one type of alkali ion by another in glasses shows that a vacancy in ζ^- after the dissociation of a large cation (for example K^+) is accessible not

---

*R. L. Myuller, Fiz. Tverd. Tela. 2(6):1345 (1960).

†γ, as in the previous papers, is equal to the ratio of the nonionized structural units $[\zeta]$ to the ionized units $[\zeta^-]$. (Editor's note).

only to large (K^+) but also to small cations (for example Li^+). Vacancies ζ^- existing after the dissociation of a small cation are inaccessible to large cations. However this does not imply the plugging up of small vacancies by large cations. The omnidirectional random intense thermal diffusion of the vacancies leads small vacancies away from the neighborhood of large cations and rapidly brings them into the neighborhood of small cations. When there is an excess of polar units with large cations the latter while drifting along the electric field, go around the small vacancies because of the thermal motion of the large vacancies (relaxation time $\tau < 2 \cdot 10^{-6}$ sec at 500°K). If there is an excess of polar units with small cations, the drift of the latter does not require them to go around the large vacancies. Where there are commensurable amounts of small and large cations they both drift according to the above description.

Introducing the new symbols $[M]_s = [M'] + [M'']$, $\beta' = [M']/[M]_s$ and $\beta'' = [M'']/[M]_s$ we can, after substituting from (2), rewrite equation (1) in the form

$$\varkappa = F[M]_s w_{0e}\left[\beta' \exp\left(-\frac{\Psi_\Phi'}{2RT}\right) + \beta'' \exp\left(-\frac{\Psi_\Phi}{2RT}\right)\right]. \tag{3}$$

Insofar as the inequalities $\beta' \gtrless \beta''$ are associated, respectively, with the inverse inequalities $\Psi_\Phi' \lessgtr \Psi_\Phi''$, one of the terms in brackets is invariably greater than the other at all temperatures, for a given β' and β''. Assuming that $\beta > \beta^*$ and correspondingly $\Psi_\Phi > \Psi_\Phi^*$ we get

$$\frac{\varkappa}{[M]_s} = F w_{0e} \beta \exp\left(-\frac{\Psi_\Phi}{2RT}\right)\left[1 + \frac{\beta^*}{\beta}\exp\left(-\frac{\Psi_\Phi^* - \Psi_\Phi}{2RT}\right)\right], \tag{4}$$

where $\log F w_{0e} = P_e$ and $\frac{\beta^*}{\beta}\exp\left(-\frac{\Psi_\Phi^* - \Psi_\Phi}{2RT}\right) = a$, and therefore

$$\log\frac{\varkappa}{[M]_s} = P_e + \log\beta - \frac{\Psi_\Phi}{4.6RT} + \log(1+a), \tag{4a}$$

where Ψ_Φ is expressed in calories.

According to our hypothesis, in glasses containing two types of alkali cations, the concentration dependence of the energy Ψ_Φ which is observed in the glasses with one type of cation, is substantially preserved for each cation. Such a concept permits us to substitute in equation (4a) the values Ψ_Φ^* and Ψ_Φ obtained from the relationships which we have established for single-cation glasses $L = \Psi_\Phi[M]^{1/2} \approx 5.0$ (borate glasses) and $L' = \Psi_\Phi[M]^{1/4} \approx 12$ (silicate glasses). We thus have a way of simplifying expression (4a).

As we established during the analysis involved in the experimental determination of L, L' and P_e, the values of Ψ_Φ for single-cation glasses have been determined by various authors to within ± 0.3 kcal. Analogously, P_e is determined to ± 0.3 [1]. It follows from this that $\Psi_\Phi^* - \Psi_\Phi$, which is ~ 3 kcal, lies within the above margin of error and need not be taken into account. For 150-300° C this corresponds to a value of $a \leq 0.3$ or $\log(1+a) \leq 0.12$, i.e., it is less than the error involved in the determination of P_e (≈ 0.3). For $\Psi_\Phi^* - \Psi_\Phi > 3$ kcal the value of $\log(1+a)$ becomes even smaller. It follows from this that we can readily neglect the value of $\log(1+a)$ in Eq. (4a).

As a result of the absence of spatial-mechanical hindrance to the translational shift of the cations to neighboring vacancies, and assuming continuous cation conduction in a medium of associated homogeneous polar groups, we can express the molar electrical conductivity of two-cation glasses in the form

$$\log\frac{\varkappa}{[M]_s} = P_e^0 + \log\beta^0 - \frac{\Psi_\Phi^0}{4.6RT}, \tag{5}$$

where Ψ_ϕ^0 is expressed in cal/mole. The zero exponent in the quantities P_e^0, β^0, and Ψ_ϕ^0 establishes, first of all, their relationship to the prevailing cation in the glass ($\beta^0 > 0.5$) and secondly, it indicates that the values of P_e^0 and Ψ_ϕ^0 in two-cation glasses are equal to the corresponding values of these quantities in single-cation glasses.

2. The empirically determined temperature dependence of the molar conductivity in two-cation glasses is expressed in the form

$$\log \frac{\varkappa}{[M]_s} = B_{10} - [M]_s - \frac{A_{10}}{T}. \tag{6}$$

A confirmation of the correctness of the proposed concept is obtained, according to equations (5) and (6), from the closeness of the values of $P_e^0 + \log \beta^0$ and $B_{10} - \log [M]_s$ or P_e^0 and $B_{10} - \log [M]_s \beta_0 = P_e$ (respectively related to the prevailing type of cation) as well as to the values of Ψ_ϕ^0 and $9.15 A_{10} = \Psi_e$.

Blocking of vacancies by the cations by the Lengyel process, or other spatial hinderances, would involve removing a significant number of the cations from the statistical distribution and the inequality $P_e \leq P_e^0$ would be true as a result of the sharp decrease in B_{10}. This, however, has not been observed. In single-cation glasses the values of P_e^0 lie within the limits 3.4-3.9 [1]. After a corresponding recalculation of the experimental data of the co-workers in our laboratory [2] we get for two-cation borate glasses the values listed in Table 1 of $P_e = B_{10} - \log [M]_s \beta^0$ between 3.2-5.0. We do not observe any decrease in the values of P_e.

On the basis of the experimental data of Mazurin and Borisovsky for the electrical conductivity of sodium potassium silicate glasses and introducing the interpolated values for the densities of glasses according to Sheybany [3], we calculated the values of P_e and $\Psi_e = 9.15 A_{10}$ given in Table 2. Thus, as for the case of borate glasses, we do not in silicate glasses observe any decrease in the values of P_e. Therefore, we may consider that the hypothesis regarding the mechanical blocking of the cations in complex glasses is not only unsound as far as theoretical considerations are concerned as we have shown previously [1], but that this hypothesis is clearly contradicted by the experimental data.

Comparing the values of P_e for the complex borate and silicate glasses in Tables 1 and 2 with the values of these quantities for simple glasses [4], it is easy to note the systematic increase in P_e for complex glasses. The increase exceeds the limits of the mean experimental deviations. We note at the same time in Tables 1 and 2 that the values of Ψ_e for complex glasses are exceeded by the values of Ψ_ϕ^0 for simple glasses, calculated using the values of L and L' for the corresponding prevailing concentrations of cations in the glass (the values of [M] which are not in brackets in Tables 1 and 2).

The associated increase in the values of P_e and Ψ_e in complex glasses corresponds to the well-known empirical compensation law [5]. An analogous correlation exists between the values of P_e and Ψ_e for glasses in the critical temperature region and is accompanied in this region by the appearance of an entropic dissociation term due to the secondary solvation of the dissociated cations [6]. We would have expected secondary solvation to occur if, during the transition from simple to complex glasses, there was a corresponding decrease in the lower limit of the critical temperature region. However, Mazurin's experimental data [7] is convincing evidence that the data shown in Table 2 refers to stabilized glasses in which there is no secondary solvation. The data of Table 1 for complex borate glasses also refers to the stabilized state [2]. The most probable reason for the concomitant increase in P_e and Ψ_e during the transition from simple to complex glasses is the regular partial dispersion of polar groupings occurring as the result of the increase in the dielectric constant of the medium on the introduction of foreign polar structural units. In fact, in this process the energy of the mutual bonds in homogeneous polar units increases and, according to Onsager—Frenkel', the interaction energy with

TABLE 1. Borate Glasses

$M'-M''$ (ξ_s)*	Glass No.	100 [M'], mole/ml	100 [M''], mole/ml	B_{10}	Ψ_e, kcal/mole	P_e	$\Delta S'$, $\frac{cal}{mole \cdot deg}$	Ψ_Φ, kcal/mole	Ψ_Φ^0, kcal/mole
Li—K ($\simeq 0.4$)	89	1.70	(0.34)	2.51	44.8	4.3	3.7	43	41
	87	1.00	(0.87)	2.59	55.5	4.3	3.7	53	54
	86	(0.72)	1.07	2.36	55.5	4.3	7.3	52	49
	85	(0.46)	1.27	2.20	51.5	4.1	5.5	49	45
	84	(0.22)	1.46	1.63	(45.6)	3.5	—	46	42
	83	(0.11)	1.56	2.44	47.2	4.2	6.4	44	41
Li—K ($\simeq 0.4$)	126	1.47	(0.20)	2.24	46.9	4.1	1.8	46	45
	125	1.25	(0.39)	2.34	49.6	4.2	2.7	48	50
	124	1.10	(0.59)	1.85	(50.4)	3.8	—	50	51
	123	0.92	(0.77)	2.50	55.3	4.5	5.5	52	56
	121	(0.60)	1.14	2.50	55.3	4.4	8.2	51	48
	120	(0.46)	1.36	1.87	49.7	3.7	1.8	49	44
	118	(0.15)	1.73	2.09	44.5	3.8	2.7	43	39
Li—Na ($\simeq 0.4$)	100	1.45	(0.49)	2.50	49.2	4.3	3.7	47	47
	99	1.01	1.01	3.28	52.7	5.0	5.5	50	49
	98	(0.49)	1.46	2.83	48.8	4.4	9.2	44	41
Na—K ($\simeq 0.5$)	133	2.02	(0.29)	1.55	(37.3)	3.2	—	37	34
	132	1.72	(0.57)	2.44	45.8	4.2	7.3	42	37
	131	1.41	(0.83)	2.90	49.8	4.6	11.0	44	41
	130	1.09	1.09	3.08	52.8	4.7	11.9	47	47
	129	(0.53)	1.59	2.82	49.0	4.6	10.1	44	40
	128	(0.20)	1.83	2.40	44.3	4.1	5.5	39	38
	127	(0.10)	1.92	2.40	42.2	4.1	5.5	37	37

* $\xi = 1/\gamma$, i.e., the ratio of concentration of ionized structural units to the non-ionized. (Editor's note).

TABLE 2. Sodium-Potassium Silicate Glasses

[Na$_2$O], mole %	[K$_2$O], mole %	d, g/cm^3	100 [Na], mole/ml	100 [K], mole/ml	100 [Si], mole/ml	B_{10}	Ψ_e, kcal/mole	P_e	ΔS, $\frac{cal}{mole \cdot deg}$	Ψ_Φ^e, kcal/mole	Ψ_Φ^0, kcal/mole
27.0	0	2.45	2.19	—	2.95	1.52	28.8	3.2	—	28.8	27.8—30.7
20.25	6.75	2.46	1.58	(0.53)	2.86	2.72	41.5	4.5	11.9	35.6	30.1—33.3
13.5	13.5	2.47	1.02	1.02	2.78	3.31	48.2	5.3	18.3	39.1	33.8—38.1
6.75	20.25	2.46	(0.50)	1.48	2.68	2.61	42.5	4.4	10.1	37.5	34.8
0	27.0	2.45	—	1.91	2.59	1.22	34.2	2.9	—	34.2	32.5
33.3	0	2.49	2.73	—	2.74	1.52	25.7	3.1	—	25.7	26.3—29.0
25.0	8.33	2.50	1.97	(0.66)	2.64	3.24	40.0	4.9	15.5	32.3	28.6—31.5
20.86	12.49	2.50	1.61	(0.97)	2.57	3.40	43.0	5.2	18.3	31.9	30.0—33.2
16.66	16.66	2.50	1.25	1.25	2.50	3.85	47.6	5.7	22.0	36.6	32.1—36.3
12.49	20.86	2.50	(0.93)	1.55	2.47	3.64	46.7	5.4	19.2	37.1	34.3
8.33	25.0	2.50	(0.57)	1.83	2.42	3.36	43.7	5.1	16.4	35.5	33.0
0	33.3	2.48	—	2.36	2.50	1.91	29.9	3.5	—	29.9	30.8
40.0	0	2.53	3.33	—	2.50	1.57	24.0	3.1	—	24.0	25.0—27.6
35.0	5.0	2.53	2.83	(0.41)	2.43	2.63	32.5	4.2	9.1	28.0	26.2—28.7
30.0	10.0	2.54	2.38	(0.80)	2.38	3.44	39.6	5.1	17.4	30.9	27.2—30.1
25.0	15.0	2.54	1.92	(1.16)	2.33	3.95	44.5	5.7	22.9	33.1	28.8—32.8
20.0	20.0	2.54	1.52	1.49	2.28	4.17	47.8	6.0	25.6	35.0	30.5—34.4
10.0	30.0	2.53	(0.72)	2.15	2.16	3.94	44.6	5.6	21.0	34.1	31.7
0	40.0	2.52	—	2.79	2.10	2.00	29.5	3.6	—	29.5	29.6

the surrounding dielectric medium increases. Such dispersion is associated with the decrease in the local concentration of dipole units which determines the decrease in the primary solvation. Thus, the more stable location of the alkali ions in the polar units in complex glasses follows naturally, and this explains the increased value of Ψ_e.

With the appearance of new polar units in the dielectric medium there is a significant increase in the number of statistically possible positions for the dissociated cations. This should be accompanied by an increase in the entropic pre-exponential term. We should also bear in mind that the differentiated association of polar structural units represents a statistical phenomenon. The partial capture of foreign dipoles and the formation of mixed groupings is inevitable [8]. The dissociation of cations in a medium with such mixed groupings also leads to an increase in the entropy term.

The entropy term $\Delta S'$, which is determined by the increased number of statistical states of the dissociated cations with the transition from simple to complex glasses, differs in principle from the entropy factor ΔS characterizing the changes in the dissociation conditions during the transition of the glass from the stabilized to the labile state. To compute the value of $\Delta S'$ we rewrite equation (6) making use of the known relationship $9.15A_{10} = \Psi_e = \Psi_\Phi^e + T\Delta S'$ where Ψ_e is an experimental directly determined quantity which includes the enthalpy associated with the dissociation of the polar units in the complex glass, Ψ_Φ^e is a quantity which has been computed with the aid of the entropic value of $\Delta S'$ and includes the free energy of dissociation in the complex glass. We then obtain

$$\log \frac{x}{[M]_s} = B_{10} - \log[M]_s - \frac{\Delta S'}{9.15} - \frac{\Psi_\Phi^e}{9.15T}. \tag{7}$$

Comparing this expression with equation (5) we get

$$\left.\begin{aligned} \Delta S' &= 9.15\left(P_e - P_e^0\right) \text{cal/mole} \cdot \text{deg}, \\ \Psi_\Phi^e \text{ (complex glass)} &= \Psi_e - T\Delta S' = \Psi_\Phi^0 \text{ (simple glass)}. \end{aligned}\right\} \tag{8}$$

The experimental values of Ψ_e and the values of $\Delta S'$ and Ψ_Φ^e for complex alkali borate and silicate glasses calculated according to (8) are shown in Tables 1 and 2. In these tables we give also the values of Ψ_Φ^0, the energy calculated with the aid of the functions L and L' for simple glasses [4] with the corresponding volume concentration of the alkali ions prevailing in the given complex glass.

Comparison of the values of Ψ_Φ^e (complex glasses) with the values of Ψ_Φ^0 (simple glasses) in Tables 1 and 2 shows that there is a sufficiently satisfactory agreement between them if we take into account the errors involved in the determination of the energy. This speaks in favor, in the case under consideration, of a regularly associated increase in P_e and Ψ_e.

3. Upon the introduction into the glass of metallic oxides with a significant polarizability (which leads to the formation in the glass of units with lower dipole moments) we naturally expect that the polar units which form should have a greater tendency toward the formation of mixed associated structures. Examining the electrical conductivity of silver thallium borate glasses which have been studied by Markin [9] it is easy to notice the unduly low values of Ψ_e with respect to Ψ_Φ^0 in these glasses, these quantities being computed from the function L of Table 3. We also note that for silver alkali glasses this occurs for $\beta_{Ag} \geq 0.5$ and for silver thallium glasses over the entire range of ratios of the cations. Given the evidently close, and at the same time small, dipole moments of the polar units B^-Ag^+ and B^-Ti^+, the thallium borate glasses are characterized by the complete absence of the minimum effect in their electrical conductivity and the absence of a maximum for Ψ_e. The negative value of $\Delta\Psi^*$ indicate the

---

* $\Delta\Psi = \Psi_e - \Psi_\Phi^0$, while $\Delta\Psi^s = \Psi_e - \Psi_\Phi^s$, where Ψ_Φ is the energy parameter determining the degree of dissociation of ions in mixed structures. (Editor's note).

TABLE 3. Borate Glasses

$M' - M''$ (ξ_8)	Glass No.	100 [M'], mole/ml	100 [M''], mole/ml	B_{10}	P_e	Ψ_e, kcal/mole	Ψ_ϕ^0, kcal/mole	Ψ_ϕ^s, kcal/mole	$\Delta\Psi$	$\Delta\Psi^s$
Li—Ag (~0.3)	183	1.42	0.19	1.73	3.6	45.2	45	—	0	—
	184	1.19	0.46	2.70	4.6	51.7	49	—	3	—
	185	0.83	0.79	2.30	4.4	49.0	56	39.3	— 7	+10
	186	0.59	1.07	1.19	3.1	37.7	48	38.8	—10	— 1
	187	0.32	1 34	1.06	2.9	34.6	43	38.8	— 8	— 4
				Mean	3.6±0.6					
K—Ag (~0.3)	167	1.48	0.15	2.29	4.1	50.2	42	—	8	—
	168	1.35	0.30	2.54	4.3	51.4	44	—	7	—
	169	0.99	0.62	2.29	4.1	50.2	51	39.4	— 1	+11
	170	0.73	0.91	1.94	3.7	45.7	52	39.0	— 6	+ 7
	171	0.50	1.15	2.25	4.0	36.5	47	38.9	—11	— 4
				Mean	4.0±0.2					
Tl—Ag (~0.2)	325	0.88	0.10	1.63	3.6	49.7	53	50.5	— 3	—0.8
	324	0.74	0.25	1.63	3.6	49.7	58	50.3	— 8	—0.6
	323	0.50	0.50	1.34	3.3	47.3	—	50.0	—	—2.7
	322	0.25	0.78	1.81	3.8	47.8	56	49.2	— 8	—1.4
				Mean	3.6±0.1					

non-applicability of the relation $\Psi_\phi^0 \cdot [M]^{1/2} = 5.0$ and the necessity, in connection with the formation of mixed structures, of introducing a different relationship: $\Psi_\phi^s ([M'] + [M''])^{1/2} = 5.0$. Actually for silver thallium glasses $\Psi_e \approx \Psi_\phi^s$, and $\Delta\Psi^s$ lies within the limits of experimental error.

In silver alkali borate glasses we note the commensurability of $\Delta\Psi$ and $\Delta\Psi^s$, which indicates that there is a development of mixed structures together with the differentiated ones.

The relatively weak variation in the value of P_e (Table 3) shows that the mobility term does not undergo any basic changes and the conductivity, during a change in composition of the glass, is determined entirely by the degree of dissociation. The latter does not change with changes in the ratio of the silver to the thallium, as a consequence of the constant number of jointly associated polar units of both types and the unchanged local primary solvation. Under these conditions equation (3) takes the form $\varkappa = F[M]_s \, \omega_{0e} \exp(-\Psi_\phi^s / 2RT)$ inasmuch as $\Psi_\phi' = \Psi_\phi'' = \Psi_\phi^s$ and $\beta' + \beta'' = 1$ in accordance with the data of Table 3.

Summary

1. We have derived the equation for the molar conductivity of glasses containing two types of univalent cations. According to this equation, when there is differentiated association of the polar structural units the electrical conductivity is determined by the flow of the prevailing type of cation in the glass.

2. Experimental data corroborates the absence of steric hindrance during the motion of the cations in complex borate and silicate glasses.

3. Experimental data on the electrical conductivity of two-cation borate and silicate glasses testify to the existence of an entropic effect during the dissociation of cations, which causes the concomitant increases in P_e and Ψ_e in accordance with the compensation law.

4. During the transition from simple single-cation glasses to complex two-cation glasses we observe a dispersion of the associated polar groupings, causing the increase in the dissociation energy of the ions and the increase in the statistical entropy term.

5. In borate glasses containing both silver and thallium polarizable cations, the mutual capture of polar structural units leads to the mixture of such units and as a result to the disappearance of the minimum conductivity effect. In this process the values of P_e and Ψ_e are independent of the relative ion concentrations.

References

1. R. L. Myuller, Zh. Tekhn. Fiz., 25:1868 (1955), ● this volume, p. 50; 26:2614 (1956), ● this volume, p. 79; Fiz. Tverd. Tela, 2(6):1333 (1960), ● this volume, p. 93.
2. B. I. Markin, Zh. Tekhn. Fiz., 10:66 (1940); N. I. Brodskaya and V. S. Tatarinova, Uch. Zap., Leningr. Gos. Univ., 54:241 (1940); L. R. Takking and N. P. Shchegoleva, Uch. Zap., Leningr. Gos. Univ., 108:17 (1949).
3. O. V. Mazurin and E. S. Borisovskii, Zh. Tekhn. Fiz., 27:275; H.-A. Sheybany, Verres et Refractaires, 3:127 (1948).
4. R. L. Myuller, Fiz. Tverd. Tela, 2(6):313 (1960).
5. P. Rutschi, Z. phys. Chemie, 14:277 (1958).
6. R. L. Myuller, Zh. Tekhn. Fiz., 25:2428 (1955), ● this volume, p. 61.
7. O. V. Mazurin, Tr. Leningr. Tekhnol. Inst., 29:72 (1954).
8. R. L. Myuller, Zh. Tekhn. Fiz., 23:1874 (1953).
9. B. I. Markin, Zh. Obshch. Khim., 16:401 (1946); Zh. Fiz. Khim. 23:1442 (1949); Fiz. Tverd. Tela, Symposium I, 214 (1959).

The Interrelation between Electrical Conductivity
and The Viscosity of Glasses

There is no direct regular physical connection of Walden's type between the electroconductivity $\varkappa$ and viscosity η of glass [1]. The known relation $\varkappa^n \eta = C$ (where n and C are constants independent of temperature [2]) can, according to Frenkel' [3], be reduced to the relation

$$ n = \frac{2E_\eta(T)}{\Psi_\Phi(T)} \neq f(T), $$

where $E_\eta(T)$ is the activation energy of flow [4] and $\Psi_\Phi(T)$ is the ion-dissociation energy, to a first approximation.

Investigations by Lark-Horowitz, Babcock and Evstrop'ev [2], Kobeko [5], Mazurin [6], and also the analysis of recent experimental data of Mazurin [7], point convincingly to the similarity in the character of the temperature dependence of $\Psi_\Phi(T)$ and $E_\eta(T)$. However, the physical nature of this similarity in the temperature dependence remains unexplained.

The following views can be formulated on the physical nature of this relationship and on the dependence of the coefficient n on the glass composition.

For lack of a direct physical relation between viscosity and the mobility of ions in glass, the similar temperature dependence $E_\eta(T) = n[\Psi_\Phi(T)/2]$ for labile glass of a particular composition is explained by a common influence, increasing with temperature, of the covalent bond switching on glass viscosity as well as on the degree of ion dissociation. The intensification of the switching of the covalent bonds produces the increasing mobility of the structural units in glass, which manifests itself in the increase of fluidity [4] and in the increase in secondary solvation of dissociated ions which lowers the ion dissociation energy [8]. The coefficient n, which is independent of temperature, is known to be determined by the glass composition. For alkali glasses, the value of n is greater than 1 and it increases with an increase in the alkali oxide content [7, 9]. This indicates a more rapid decrease in Ψ_Φ than in E_η when the degree of ionization ξ of the covalent glass network increases. The latter, in our opinion, can be explained by the fact that the fluidity activation energy E_η decreases with an increase in the degree of ionization ξ of the network due only to an intensification of the covalent bond switching. The lowering of the energy Ψ_Φ for the same increase in ξ is caused, not only by the intensification of the covalent bond switching which determines the increase of the secondary solvation of dissociated ions, but also by a rise in the primary solvation caused by the increase in the concentration of polar structural units.

*R. L. Myuller, Fiz. Tverd. Tela, 1(2):346 (1959).

In the case of alkaline borate glasses, it is possible to derive a relation between the fluidity activation energy $E_\eta(T)$ and the energy $\Psi_\Phi(T)$ based on the experimental data of Chartsis and collaborators [9],

$$E_\eta(T) = (0.325 + 3.00\xi + 1.65\xi^2)\,\Psi_\Phi(T).$$

REFERENCES

1. P. Walden, Das Leitvermögen der Lösungen, Part III (Leipzig, 1924); R. L. Myuller, Zh. Tekhn. Fiz. 25:236 (1955), ● this volume, p. 15.
2. K. Lark-Horowitz and C. Babcock, Phys. Rev., 44:321 (1933); J. T. Littleton, Ind. Eng. Chem., 25:748 (1933); K. S. Evstrop'ev, Zh. Fiz. Khim., 6:454 (1935).
3. Ya. I. Frenkel', Kinetic Theory of Liquids, Izd. Akad. Nauk SSSR, Moscow (1945).
4. R. L. Myuller, Zh. Prikl. Khim., 28:363, 1077 (1955).
5. P. P. Kobeko, Amorphous Substances, Izd. Akad. Nauk SSSR, Moscow (1952).
6. O. V. Mazurin, Tr. Leningr. Tekhnol. Inst., 29:72 (1954).
7. O. V. Mazurin, Izv. Vysshikh Uchebn. Zavedenii Khim. i Khimich. Tekhnol., 1(3):136 (1958).
8. R. L. Myuller, Zh. Tekhn. Fiz., 25:2428 (1955), ● this volume, p. 61.
9. L. Chartsis, W. Capps, and S. Spinner, J. Am. Ceram. Soc., 36:319 (1953).

The Ionic Conductivity of Alkali Aluminosilicate Glasses[*]

Aluminosilicates in a glassy form are frequently used as materials in the radio and electrical industry and therefore the effect of aluminates on the properties of dielectrics is of interest. The introduction of Al_2O_3 into alkali silicate glasses is accompanied by a significant increase in the dielectric constant, dielectric loss and electrical conductivity, in contrast to the introduction of borates which leads to a decrease in the conductivity and loss while the dielectric constant is increased [1-6].

On melting an alkali silicate glass with alumina, the alkali oxide and the aluminum oxide come together

$$Na_2^+O^{2-} + Al_2O_3 \rightarrow 2Na^+Al^-O_{4\,2} - 48 \text{ kcal,}$$

and a new covalent bond, $Al-O$, appears and fixes the stable $Na^+Al^-O_{4/2}$ structural unit.

The increased dielectric loss in alkali aluminosilicates is explained by the occurance of a relaxation displacement of bonded cations in the $Al^-O_{4/2}$ group

The activations energy for such a displacement, like the conductivity determining energy of electrolytic dissociation of the polar $M^+Al^-O_{4/2}$ groups, is less than that for $M^+O^-SiO_{3/2}$ and $M^+B^-O_{4/2}$ groups. This is in keeping with the successively smaller distance between the charge ceners in these groups: $r(M^+Al^-) > r(M^+O^-) > r(M^+B^-)$ [7,10]. It is well known that the thermal energy expended in removing an M^+ cation from Al^-, O^-, or B^- and in overcoming the repulsion of the electron shells of the atoms is compensated, to a considerable extent, by the solvation energy which arises from the interaction between the cation set free from the polar group and the surrounding dipolar groups. The latter become oriented by the local displacements of bonded cations as shown in Eq. (1).

When alumina is added to alkali aluminosilicate glass the content of the $M^+Al^-O_{4/2}$ polar increases and on statistical average the displacement of alkali cations is facilitated, leading to an increase in the dielectric constant, the conductivity and the loss [1-6,10]. This phenomenon continues until all the alkali metal oxide in the glass has been absorbed by the added alumina i.e. until $\{ [Al]/[M^+] = 1 \}$. A further addition of alumina results in the accumulation of nonpolar groups which produce the highest packing density (and not loose packing as Grechanik suggests in [3]) of the oxygen atoms (in quartz the density of the $[SiO_{4/2}]$ structural units ≈ 0.04 mole/cm³; in boric anhydride $[BO_{3/2}] \approx 0.05$ mole/cm³; in various modifications of alumina $[AlO_{3/2}] \approx 0.07$ mole/cm³). This condensation determines the energy expended by the

*R. L. Myuller and A. A. Pronkin, Zh. Prikl. Khim., 36(6):1192 (1963).

cations in overcoming the repulsive forces of the electron shells of the atoms in the network when the $M^+Al^-O_{4/2}$ polar group dissociates. This explains the increase in the energy $\varepsilon_\sigma{}^*$ when $[Al]/[M^+] > 1$ [10] and the decrease in conductivity

$$\sigma = \sigma_0 \cdot \exp\left(-\frac{\varepsilon_\sigma}{2kT}\right) \tag{2}$$

in alkali silicate glasses. A similar decrease of the electrical conductivity is sometimes further enhanced by the decrease in the concentration of polar groups in the glass resulting from substitution of one $SiO_{4/2}$ group by two $AlO_{4/2}$ groups [5, 6]. An analogous increase in the dielectric constant is observed when boric anydride is introduced into alkali silicate glasses (the orientation of $M^+B^-O_{4/2}$ dipoles is easier than the orientation of $\overset{\diagdown}{\diagup}Si-O^-M^+$ quadrupoles, $M^+O^--Si\overset{\diagup}{\diagdown}$

but the conductivity and dielectric loss fall in this case because $r(M^+O^-) > r(M^+B^-)$. In order to understand correctly the electrical properties of the more complicated alkali alumino-borosilicate glasses [6] we require a meticulous preliminary analysis of the atomic-ionic volume concentration relationship between the structural elements in such glasses.

The above picture allows us to state with confidence that the "anomalies" in the electrical properties of alkali aluminosilicate glasses cited in the literature can now be satisfactorily explained by an ionic-atomic mechanism for these phenomena in glasses. The deformation losses with high frequency resonance maxima in the temperature range 40-70°K, are explained by resonance deformation oscillations in an alternating field of the covalently (but not homopolar [2]) bonded atoms in the glass network [13-15]. These oscillations have energies < 0.1 eV, which according to the thermal capacity data [16] correspond to deformation oscillations excited at low temperatures.

An electron-resonance mechanism is also proposed for the dielectric loss and conductivity. This concept attributes the dielectric loss in alkali aluminosilicate glasses in the 200-500°K range to a resonance displacement of electrons in the tetrahedral $Al^-O_{4/2}$ group [4, 17-19];

$$\tag{3}$$

and suggests that over this same temperature range there is mixed electron-ionic conductivity [17].

The similarity between the stable, saturated electron-octet, tetrahedral structures $Al^-O_{4/2}$ and $SiO_{4/2}$ which, in contradistinction to the defect and irradiated structures considered in [20], do not have electron vacancies as considerable energy would be required to break the valence bonds of the coordinating atoms, and the absence of increased losses in alkali-free feldspar [21, 22], is evidence against the electron resonance origin of the dielectric loss in question. The resonance frequency ($\log f \cong 5$-6 at 300°K [17]) corresponding to an energy of 0.4 eV, is in good agreement with the energy of the potential barrier to displacement of the M^+ cation in the polar, tetrahedral structural units [7].

As well as the absence of any basis for the assumption of an electron-resonance origin of dielectric loss there is no real basis either for assuming mixed electron-ion conduction in alkali aluminosilicate glasses.

*In the preceding papers in this series R. L. Myuller uses the symbol Ψ_Φ for this parameter. (Editor's Note)

Individual glasses (Nos. 3 and 15 in the data of Ioffe and Khvostenko [19]) distinguished by the bends in the curves of log $\sigma(1/T)$ and with low energy conductivity have, below 80 and 230°C, conductivity moduli [23] which correspond to only 1 in 10^6–10^{16} of the $Al^-O_{4/2}$ tetrahedra present in the glasses.

Such are, clearly, only apparent conductivities and are due to the slow establishment and disappearance (reported in [19]) of high voltage polarization and to the slow drift of the induced charges.

We undertook this present study in order to obtain direct experimental evidence of mixed electron–ion conductivity in aluminosilicate glasses over the temperature range 200–500°K.

If, according to [19], there is a transition from electron to ion conduction in aluminosilicate glasses over this temperature range then it must be accompanied by considerable changes in the values of σ_0 and ε_σ in the specific conductivity expression, Eq. (2). The transition from electron conduction in the glass at low temperatures with an energy $\varepsilon_{\sigma e}^{low\ T}$ to high temperature ionic conductivity with an energy $\varepsilon_{\sigma_i}^{high\ T}$ is only possible when $\varepsilon_{\sigma e}^{low\ T} \neq \varepsilon_{\sigma_i}^{high\ T}$. We must remember that according to both concepts the source of the current carriers, M^+ ions and electrons, is the structural element $M^+Al^-O_{4/2}$. When the concentration of $[M^+Al^-O_{4/2}] = n$ mole/cm³, the theoretical value of the conductivity modulus $\log \sigma_{0\,t}/n = 4 \pm 1$ for both ion and electron conduction [23]. But in the temperature range where the electron conduction changes to ionic there must be a conductivity entropy $\Delta S_\sigma = -\Delta \varepsilon_\sigma / \Delta T$ and a large experimental value will be observed for $\sigma_{0e} = \sigma_{0t} \exp\left(\dfrac{\Delta S_\sigma}{2k}\right)$ [24]. The conductivity modulus is

$$\log \frac{\sigma_{0e}}{n} = \log \frac{\sigma_{0\,t} \exp\left(\Delta S_\sigma/2k\right)}{n} \gg 4.$$

In the opposite case, i.e., when the value of the modulus $\log (\sigma_{0e}/n) = 4 \pm 1$ is invariable, it is necessary in this particular temperature range that $\varepsilon_\sigma = $ const, which thus eliminates the possibility of a transition from one type of conductivity to another. In this case the existence of a high temperature ionic conduction would indicate an invariant ion-type conductivity over the entire temperature range in question for aluminosilicate glasses.

The results of the study presented below of the electrical conductivity of sodium aluminosilicate glasses in the transition temperature region and of the transport number of the sodium ion at high temperatures, confirms that in these glasses such conduction is purely ionic.

The experimental data of Ioffe and Khvostenko [19] for the temperature dependence of the conductivity of aluminosilicate glasses cannot be used to calculate the conductivity modulus since in this work the charge and discharge currents were extrapolated to zero time. It was assumed that the fall in current was represented by a uniform function with continuous derivatives. The inadmissibility of a similar extrapolation was noted previously in [25]. When the initial current as well as the transportive current of free cations was similarly calculated, the relaxation displacement of the bonded cations was allowed for, the lower the temperature the greater the extent. From such calculations, therefore, we obtain an overestimate of the low temperature conductivity and from a graphical treatment of the experimental data for the log $\sigma(1/T)$ dependence we obtain values for ε_σ and log σ_0 which are low compared with the values obtained by a strict determination of the steady-state transportive current with an active anode.

The glasses were melted in quartz crucibles at 1450–1500°C for 6 to 8 h and stirred with an iron rod. The melt was poured out on a steel table and the glass briquets so obtained were placed in a muffle and kept at 600°C for 30 min. The briquets were then allowed to cool at the natural rate of the muffle over 12 h. These briquets were sliced and the specimens polished to produce plane-parallel (± 0.01 mm) discs, 20 mm in diameter and about 2 mm thick. The chemical compositions by synthesis of the experimental glasses Nos. 1–7 are given in Table 1 which also shows the average values of the density, determined on the parallel faced specimens by suspension in toluene.

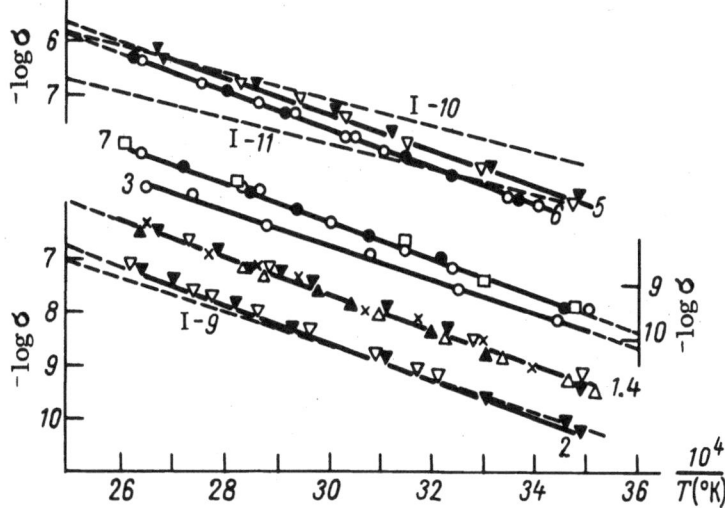

Fig. 1. Dependence of the logarithm of specific conductivity log σ of sodium aluminosilicate glasses on the reciprocal of temperature: 1-7) the numbers of the glass in Table 1; different specimens of glass of the same composition are denoted by different symbols.

The microhardness values shown in Table 1 were obtained with a PMT-3 instrument. The steady state electrical conductivity was measured by a conventional galvanometric method and checked by a potentiometric instrument. Sodium amalgam was used as the anode. In glasses Nos. 1, 2 and 6 the cathode was also sodium amalgam while in glasses Nos. 2-5, and 7 a copper cathode was used. It is clear from the measurements on the parallel specimens of No. 2 that they gave the same results (see Fig. 1).

The experimental results for the temperature dependence of conductivity in the glasses we investigated are given in Fig. 1. The dotted lines show the temperature dependence of the conductivity of glasses Nos. I-9, I-10, and I-11 reported by Ioffe and Khvostenko in [19]. A comparison of the latter results with ours for glasses Nos. 2, 5, and 6 which were close in composition, and a similar comparison of the values of ε_σ and log σ_0 for these glasses in Table 1 confirms the underestimation of these values by Ioffe and Khvostenko [19]. The greatest divergence, between glass No. I-11 and No. 6, is explained by the probability of anode polarization when nonactive anodes are used [19].

A fundamental result of our study of the conductivity of sodium silicate glasses has been to establish the constancy of the modulus of electrical conductivity in the "transition" region 15-100°C where log (σ_0/n) = 4.2 ± 0.3 (Table 1), in fairly good agreement with the theoretically calculated values (3.7 ± 1, in the case of ion conduction and 4.6 ± 1, for electron conduction [23]). Here n = $[Na^+Al^-O_{4/2}]$ = $\varkappa[M]$ mole/cm³, where [M] = d/M is the concentration of the average chemical unit M = $(Na^+Al^-O_{4/2})_x \cdot (Na^+O-SiO_{3/2})_{1-x}(SiO_{4/2})_y$ (Table 1).

The constancy of the value of the modulus and the correspondence between the numerical and the theoretical value, point to the absence of a change in the nature of the current carriers over this "transition" range of temperature. The nature of the current carriers was established by determining the transport number.

In order to determine the transport number for the sodium ion by the Tubandt method, three plane-parallel glass specimens of a given composition were clamped between two silver electrodes (10 mm diameter) and placed in a cylindrical heat-resistant steel furnace with a brass

TABLE 1. Modulus of Electrical Conductivity in Sodium Aluminosilicate Glasses

Glass No. (No. of specimens)	Composition of glass						Mole wt. of M	Density d, g/cm³	Conc. [M]·10², mole/cm³	Micro-hardness H, kg/mm²	ε_σ', eV	$\log \sigma_0$	$\log \dfrac{\sigma_0}{x[M]}$
	Na_2O	Al_2O_3	SiO_2	Structural chemical unit M									
	mole %			x	$1-x$	y							
1 (1)	30.0	10.0	60.0	0.33	0.67	0.33	108.0	2.47₄	2.28	566±15	1.33	2.35	4.5
2 (2)	20.0	10.0	70.0	0.50	0.50	1.25	161.6	2.41	1.50	580±20	1.41	2.00	4.1
I-9	20.6	9.4	70.0	0.46	0.54	1.16	—	—	—	—	1.32	1.20	—
3 (1)	25.0	10.0	65.0	0.40	0.60	0.70	129.3	2.44₆	1.90	560±14	1.25	1.13	3.3
4 (4)	30.0	20.0	50.0	0.67	0.33	0.49	114.6	2.46₃	2.15	568±12	1.35	2.5	4.3
5 (2)	25.0	20.0	55.0	0.80	0.20	0.90	137.7	2.45₄	1.78	647±15	1.30	2.37	4.2
I-10	25.0	21.3	53.7	0.85	0.15	0.93	—	—	—	—	1.01	0.39	—
6 (2)	25.0	15.0	60.0	0.60	0.40	0.80	133.6	2.45₂	1.84	—	1.37	2.71	4.7
I-11	25.0	16.3	59.8	0.65	0.35	0.85	—	—	—	—	1.13	0.77	—
7 (4)	30.0	15.0	55.0	0.50	0.50	0.42	111.5	2.48	2.22	565±20	1.37	2.4	4.4

Mean . . . 4.2 ± 0.3

cover. The glass and electrodes were clamped by a spring and screw between an optical quartz plate and an optical quartz rod clamped elastically to the silver cathode. The temperature was measured by a thermocouple in contact with the silver cathode. The resistance of the insulation was $\geq 10^{12}\ \Omega$. The temperature in the steel cylinder was automatically controlled to an accuracy of $\pm\ 0.1°$ using a differential photoresistance and a thermocouple enclosed in the steel wall of the furnace. The voltage at the electrodes was stabilized within $\pm\ 0.5\%$ and was supplied from a universal power supply (UIP-1) using a ballast resistance of $10^4\ \Omega$. The quantity of electricity flowing through the glass specimens was measured to $\pm\ 1.5\%$ by a standardized X-11 type electrolytic coulombmeter. The individual glass discs were weighed with an accuracy of $\pm\ 0.0001$ g.

After about 20 C had been passed through the glass packs the current was switched off and the furnace and specimens were cooled over 10 to 12 h to room temperature.

The glass, enriched by the silver from the anode side, was colored yellow. When more than 30 C had been passed, cracks were found in the anode glass and as a result it was perhaps somewhat pulverized (see $\eta_{Na}^{\text{anod. glass}}$ for Glass No. 2, Table 2).

The calculation of the transport number η_{Na} from Tubandt's data [26] was based on coulometer readings (Q is the equivalent of the amount of electricity flowing) and on changes in weight of the silver anode (ΔAg is the number of equivalents of silver replacing the departed sodium ions in the glass), and of the anode glass [$\Delta (Ag-Na)$ is the equivalents of the sodium ions which have left the glass and are replaced by silver ions] expressed in equivalents:

$$\eta_{Na}^{\text{anode}} = -\frac{\Delta Ag}{Q} \quad \text{and} \quad \eta_{Na}^{\text{anod. glass}} = \frac{\Delta (Ag - Na)}{Q}. \tag{4}$$

The experimental data and the results of the calculation of the transport numbers of the sodium ions in the sodium aluminosilicate glasses Nos. 1–7 used in this work are shown in Table 2 and their composition is shown in Table 1.

TABLE 2. The Transport Number for the Na Ion in Aluminosilicate Glasses

Glass No.	Temperature, °C	Quantity of charge Q		Silver anode				Anode glass				Average glass			Sodium ion transport number	
		coulomb	g-eq $\cdot 10^4$	wt. before exp., g	wt. after exp., g	decrease in wt. $\cdot 10^2$, g	decrease in g-eq $-\Delta Ag \cdot 10^4$	wt. before exp., g	wt. after exp., g	increase in wt. $\cdot 10^2$, g	increase in g-eq $\Delta(Ag-Na) \cdot 10^4$	wt. after exp., g	wt. before exp., g	change in wt. $\cdot 10^2$, g	anode η_{Na}	anode glass η_{Na}
1	301.2	21.9	2.27	3.8147	3.7902	2.45	2.27	1.4617	1.4807	1.90	2.24	1.4530	1.4529	−0,01	1.000	0.987
1′	296.4	21.4	2.22	3.7624	3.7384	2.40	2.23	1.4633	1.4820	1.87	2.21.	1.4506	1.4505	−0.01	1.002	0.995
2	259.0	35.3	3.64	3.7260	3.6869	3.91	3.62	1.4082	1.4370	2.88	3.39	1.4132	1.4134	+0.02	0.995	(0.935?)
2′	293.4	21.4	2.22	3.9421	3.9182	2.39	2.21	1.4161	1.4347	1.86	2.19	1.4262	1.4263	+0.01	0.995	0.985
3	291.3	22.0	2.30	3.7895	3.7638	2.57	2.29	1.4470	1.4660	1.90	2.24	1.4414	1.4413	−0.01	0.995	0.975
4	301.2	21.8	2.26	3.8679	3.8436	2.43	2.25	1.4649	1.4838	1.89	2.23	1.4593	1.4593	0.00	0.995	0.987
5	301.2	19.9	2.06	3.8909	3.8687	2.22	2.06	1.4483	1.4656	1.72	2.02	1.4309	1.4307	−0.02	1.000	0.982
6	290.0	49.7	5.17	4.0404	3.9850	5.54	5.12	1.4473	1.4898	4.25	5.01	1.4408	1.4411	+0.03	0.990	0.970
6′	289.6	22.3	2.31	4.1350	4.1106	2.44	2.27	1.4374	1.4567	1.93	2.27	1.4402	1.4402	0.00	0.990	0.970
7	290.0	19.9	2.06	3.9688	3.9408	2.20	2.04	1.4503	1.4672	1.69	1.99	1.4488	1.4486	−0.02	0.990	0.992
7′	294.1	21.5	2.23	3.9163	3.8920	2.43	2.25	1.4473	1.4659	1.86	2.21	1.4511	1.4511	0.00	1.008	0.982
														Mean	0.996 ± 0.005	0.983 ± 0.006

It is clear that cation conduction was observed in the range 260–300°C in the glasses in question.

This result, together with the relationship between the experimental and the theoretically calculated modulus of electrical conductivity established earlier, unequivocally confirms the purely ionic character of conduction in the range 300–570°K in the aluminosilicate glasses under examination.*

Conclusions

1. The "anomalous" electrical properties of alkali aluminoborosilicate glasses have been satisfactorily explained in terms of the particular feature of the transport of free alkali cations and of localized displacements of bonded cations in the polar units of the glass network.

2. There are fundamental objections to the electron-resonance concept of dielectric loss in alkali aluminosilicate glasses and this concept cannot be considered as valid.

3. The experimental study of the temperature dependence of electrical conductivity of a series of sodium aluminosilicate glasses in the 15–100°C range established the constancy of the modulus log $(\sigma_0/n) = 4.2 \pm 0.3$, and its good agreement with the numerical values from the theoretical calculations.

4. The ionic conductivity of sodium aluminosilicate glasses from 260–300°C was established with an accuracy of 0.5% using the Tubandt method for determining the sodium ion transport number.

5. The experimental values for the conductivity modulus in the temperature interval 15–110°C, and the sodium ion transport number from 260–300°C, established the purely ionic character of the conductivity of sodium aluminosilicate glasses from 15–300°C.

*It was subsequently proved experimentally by Pronkin that the transport number of sodium in aluminosilicate glasses is unity.

References

1. L. Navias and K. L. Green, J. Amer. Ceram. Soc., 29:267 (1946).
2. N. M. Verebeichik, V. I. Odelevskii, and L. É. Kamenchik, Zh. Tekhn. Fiz., 22:12 (1952).
3. L. A. Grechanik, Tsentr. Nauch. Issled. Lab. Élektrotekh. Stekla, 1:3 (1954); Zh. Prikl. Khim., 31:1164 (1958).
4. V. A. Ioffe, Zh. Tekhn. Fiz., 24:661 (1954); 26:516 (1956); 27:1453 (1957).
5. N. M. Verebeichik and V. I. Odelevskii, Zh. Tekhn. Fiz., 26:1696, 1704 (1956).
6. A. A. Appen and Kan Fu-hsi, Fiz. Tverd. Tela, 1:1529 (1959); in: The Structure of Glass, Consultants Bureau, New York (1960) p. 445.
7. R. L. Myuller, Zh. Tekhn. Fiz., 25:1556 (1955), ● this volume, p. 33.
8. V. I. Odelevskii and N. M. Verebeichik, Izv. Tomsk. Politekhn. Inst., 91:247 (1956).
9. R. J. Charles, J. Appl. Phys., 32:1115 (1961).
10. J. O. Isard, J. Soc. Glass Technol., 43:113 (1959).
11. O. L. Anderson and D. A. Stuart, J. Amer. Ceram. Soc., 37:573 (1954).
12. E. Mooser and W. B. Pearson, Nature, 190:406 (1961).
13. J. M. Stevels, J. Soc. Glass Technol., 34:80 (1950).
14. V. A. Ioffe, Dokl. Akad. Nauk SSSR, 87:405 (1952).
15. J. Volger and J. M. Stevels, Philips Res. Repts., 11:452 (1956).
16. R. L. Myuller, in: The Structure of Glass, Vol. 2, Consultants Bureau, New York (1960), p. 50, ● this volume, p. 178.
17. V. A. Ioffe and I. S. Yanchevskaya, Zh. Tekhn. Fiz., 28:2154 (1958); in: Dielectric Physics, Izd. Akad. Nauk SSSR, Moscow (1960), pp. 182 and 218; Fiz. Tverd. Tela, 4:676 (1962).
18. V. A. Ioffe, G. I. Khvostenko, and I. S. Yanchevskaya, in: The Structure of Glass, Vol. 2, Consultants Bureau, New York (1960), p. 244.
19. V. A. Ioffe and G. I. Khvostenko, Fiz. Tverd. Tela, 2:509 (1960).
20. J. Susmann, Tech. Rep. Electr. Res. Assoc., 348 (1956).
21. V. I. Odelevskii, N. M. Verebeichik and L. M. Ped'ko, in: Dielectric Physics, Izd. Akad. Nauk SSSR; Moscow (1960), pp. 170 and 217.
22. V. I. Odelevskii and A. F. Khomylev, in: The Structure of Glass, Consultants Bureau, New York (1960), p. 250.
23. R. L. Myuller, Zh. Prikl. Khim., 35:541 (1962), ● this volume, p. 121.
24. R. L. Myuller, Zh. Fiz. Khim., 6:616 (1935); Zh. Tekhn. Fiz., 25:236, 246, 2428 (1955), ● this volume, pp. 15, 24, 61; Fiz. Tverd. Tela, 2:1345 (1960); R. Myuller (Müller), Acta Physicochim. URSS, 2:103 (1935).
25. H. Schiller, Z. Phys., 42:246 (1927); 50, 577 (1928).
26. G. Ostroumov, Zh. Obshch. Khim., 19:363 (1949).

PART 2

THE ELECTRICAL CONDUCTIVITY
OF VITREOUS SEMICONDUCTORS

The Nature of the Electrical Conductivity
of Vitreous Semiconductors

The band theory of conduction is based on the assumption of a perfect crystalline structure of solids. Experimental investigations of the conductivity of liquids and amorphous bodies have shown that this condition is not always satisfied [1]. The fact that long-range order is not essential, and that the electron energy spectrum remains unchanged on melting if short-range order is retained, has now conferred primary importance on problems relating to structural coordination numbers and the nature of the chemical bonds in semiconductors [2, 3]. Even at the time when Wilson first postulated a band theory [4], LeBlanc had suggested that there was a correlation between electronic conduction and the covalent nature of the chemical bond and had noted that chemical compounds with well-defined polar-ionic bonds in the lattice exhibit ionic conduction [5].

Recently, continuing attempts at a mathematical analysis within the framework of the band theory of the nature of the conductivity of various chemical substances have been accompanied by investigations of semiconducting properties based on concepts of short-range order and localized chemical valence bonds [3, 6]. In continuation of this work, an attempt is made in this paper to calculate the value of σ_0 in the well known expression for intrinsic conductivity:

$$\sigma = \sigma_0 \exp\left(-\frac{\varepsilon}{2kT}\right) \tag{1}$$

for covalently bonded semiconductors with low mobility carriers.

In the very simple covalently bonded polymeric selenium chain $[-Se-]_n$ successive pairs of atoms are linked by homopolar electron-pair bonds. At the present time there is no clear idea of the structure of such electron bonds in a solid. Consideration of such bonds is confined to a somewhat arbitrary use of wave-mechanical probability functions for the distribution of the valence electrons of individual atoms.

In the examination of the dynamics of motion of carriers in polymeric solids which follows, a kinematic model is postulated (Fig. 1; linear projection of the selenium chain). The polymeric chain of selenium atoms is represented as consisting of links in the form of two-centered electron clouds which discretely bond neighboring atoms in pairs. The bonding electron cloud, very rarefield at the "perigee" at each of two Se^{++} ions, is sharply condensed in the interionic "apogee." The two electron-pair clouds immediate to the Se^{++} ion are mutually discrete and do not coalesce, as they lie in different planes. Frenkel', who opposed the unrestricted application of formal band concepts, considered the analogous bond of electrons with a pair of neighboring at-

*R. L. Myuller Zh. Prikl. Khim. 35:541 (1962).

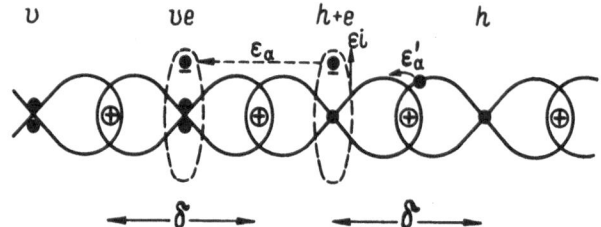

Fig. 1. Model of a polymeric chain of atoms with discretely localized electron bonds (linear projection of a selenium chain).

oms to be probable [7]. The model put forward here in projection can be regarded as adequate for the overlap of characteristic atomic functions [8]. The author used similar discrete localized electron bonds as the basis for a treatment of thermal capacity and fluidity of solids with structural networks and for an analysis of devitrification [9].

During the redistribution of the thermal vibrational energy the individual discrete interatomic electron-pair valence bonds (v) periodically absorb energy quanta ε_i with the transition of one of the two electrons of the bond to a local excited orbital. This gives rise to a weaker single-electron bond which may be regarded as a vacancy or hole (h) with an effective positive charge. Thus, a two-electron v-bond is converted into a single-electron h-bond and an electron in an excited orbit, or $v \rightarrow h + e$ (Fig. 1).

The excited electron and hole, overcoming the new slight activation barrier ε_a, are displaced in different directions and ionization of the electron-pair bond is thereby completed. The excited electron is arrested briefly at the individual links between atoms, (ev), retained by weak polarization forces. When an excited electron meets a hole they may recombine. Accordingly, the equilibrium equation for the reversible reaction of ionization of an electron-pair bond may be written as follows:

$$2v + \varepsilon_i \rightleftarrows h + ev.$$

The ionization constant of the electron-pair bond may be expressed in the form

$$\frac{n_h n_e}{n_v^2} = K = \frac{Q_h Q_e}{Q_v^2} e^{-\frac{\varepsilon_i}{kT}}, \tag{2}$$

where n_h, n_e, and n_v are the concentrations and Q_h, Q_e and Q_v are the sums over states of holes, excited electrons, and electron-pair valence bonds.

In the case of intrinsic conduction we have

$$n_h = n_e = n_i \ll n_v.$$

Following the approximation adopted for semiconductors (for the condition $\varepsilon_i \gg kT$) the Fermi−Dirac statistics are reduced to the Maxwell−Boltzmann quantum statistics. The very low degree of ionization of covalent bonds in semiconductors permits the application of the law of mass action in terms of concentration without the use of activities. The sum over states Q_v of the electron-pair in the initial normal state is unity. The sum over states Q_e of the electron in the excited interatomic orbital is two, as is the sum over states Q_h of an electron remaining in the normal state. Equation (2) can therefore be rewritten in the form

$$n_l = 2 n_v e^{-\frac{\varepsilon_i}{2kT}},$$

(3)

where n_v is the number of covalent bonds in cm^{-3}.

In accordance with our model there are no grounds for assuming appreciable exchange of electrons between neighboring discrete electron-pair clouds (v). However, it is reasonable to postulate electron transfer from a saturated electron-pair cloud (v) to a neighboring single-electron hole (h) in the form of a small nonwave displacement of the electron at the perigee by surmounting the low potential barrier ε_a:

$$v\,(1) + h\,(2) \rightleftarrows h\,(1) + v\,(2).$$

Analogously, excited electrons with weak polar bonding may be displaced with low activation energy $\varepsilon_a^!$ (Fig. 1):

$$ve + v \rightleftarrows v + ev.$$

Random thermally activated diffusion of holes and electrons will occur in this way. The atomic vibrational frequency, which we henceforth assume to be $\nu = 10^{13}$ sec^{-1} to an order of magnitude, plays a statistically significant part here. Hence, the number of activated displacements of electrons or holes per second per unit volume is

$$n_l \nu \exp\left(-\frac{\varepsilon_a}{kT}\right).$$

The very slight actual shift of the electron at the perigee is accompanied by an effective shift of the electron cloud by a distance corresponding to the interatomic spacing δ (Fig. 1); in the cases considered below it lies within the limits $2 \cdot 10^{-8} \le \delta \le 4 \cdot 10^{-8}$ cm. Application of an electric field E causes drift of the charge carriers. The number of carriers of a given type passing through 1 cm^2 per second along the field is

$$\vec{n_l} = \frac{\nu \delta n_l}{3}\left[\exp\left(-\frac{\varepsilon_a - 0.5\delta eE}{kT}\right) - \exp\left(-\frac{\varepsilon_a + 0.5\delta eE}{kT}\right)\right].$$

Since $\delta eE \ll kT$, we have

$$\vec{n_l} = \frac{\nu \delta^2 n_l eE}{3kT} \exp\left(-\frac{\varepsilon_a}{kT}\right).$$

(4)

Hence the mobility of the charge carriers is

where

$$u_l = \frac{\vec{n_l}}{n_l E} = \frac{\nu \delta^2 e}{3kT} \exp\left(-\frac{\varepsilon_a}{kT}\right) = u_0 \exp\left(-\frac{\varepsilon_a}{kT}\right),$$

$$u_0 = \frac{\nu \delta^2 e}{9 \cdot 10^2 kT} = 3.87 \cdot 10^3 \frac{\nu \delta^2}{T} \ cm^2/V \cdot sec.$$

(5)

Taking into account the presence of two types of carriers and assuming that $u = u_e + u_h \approx u_i$, we find, after substitution of Eq. (3), an expression for specific conductivity:

$$\sigma = 2 u_l e n_l = 4 u_0 e n_v \exp\left(-\frac{\varepsilon_i + 2\varepsilon_a}{2kT}\right),$$

(6)

or, in accordance with Eqs. (1) and (5):

$$\varepsilon = \varepsilon_i + 2\varepsilon_a,$$
$$\sigma_0 = 4u_0 e n_v = 4u_0 F [v] = 3.9 \cdot 10^5 u_0 [v] \ \Omega^{-1} \cdot cm^{-1}, \qquad (7)$$

where F = 96540 C and (v) is the concentration of valence bonds (mole/cm³).

At an average temperature of 50–100°C (about 350°K), a frequency $\nu \approx 10^{13} sec^{-1}$, and an average value $\delta \approx 3 \cdot 10^{-8}$ cm we have, in accordance with Eqs. (5) and (7)

or

$$\sigma_0 = 4 \cdot 10^4 [v] \ \Omega^{-1} \cdot cm^{-1}.$$
$$\log \frac{\sigma_0}{[v]} = 4.6. \qquad (8)$$

In accordance with our model, Eqs. (6) and (8) take into account both activated displacements of holes within the electron bond framework and displacements of electrons in excited orbitals adjacent to the links of the framework and penetrating it in its gaps.

If the continuity of the bond framework is broken by cracks and and cavities in the semiconductor, the experimental value of σ_{0e} will be lower than σ_0 calculated from Eq. (8) (case I). A similar lowering of the experimental σ_{0e} in comparison with the calculated value will be observed as the result of appreciable blocking of carriers by the conversion of a considerable proportion of the continuous chain structures into closed loops and maze-like structures with dead ends, such as are found in selenium (case II, [10]).

For a quantitative estimate of this deviation of the conductivity from the calculated value (8) we introduce a steric factor β:

$$\sigma_{0e} = \beta \sigma_0 = 4.0 \cdot 10^4 \beta [v] \ \Omega^{-1} \cdot cm^{-1}. \qquad (9)$$

In cases I and II above the values of β will be less than unity if the values of activation energy ε_a remain unchanged.

According to Hippel, blocking of this type is possible in ring structures (for example, in crystalline red selenium) in relation to holes [11]. Insofar as we assume that the electron orbits are adjacent to the links in the framework, it is reasonable to assume that holes and electrons are blocked in cyclic structures to an equal degree.

In the critical temperature region in a covalently bonded solid the current carriers may escape a 'blockage' by displacements from a link in one ring to a link in another at the instant of overlap of the electron clouds during the rotational–vibrational motions of individual kinematic pairs, as in the viscous flow of polymeric glasses [9]. For example, a single-electron bond may be transferred from one ring to another in this manner (Fig. 2). The activation energy of the overlap of the electron orbitals corresponds to the energy of transformation of valence bonds ε_a in viscous flow of polymeric materials; here in case III $\varepsilon_t \gg \varepsilon_a$ [9], and by Eq. (7) we have $\varepsilon = \varepsilon_i + 2\varepsilon_t$. Here the activation entropy of bond switching ΔS determines the value of the coefficient $\beta = \exp(\Delta S / 2k) > 1$ [12-14].

At sufficiently high temperatures in a molten polymeric semiconductor an entropic switching of valence bonds takes place (case IV; $\Delta S = 0$, $\beta \approx 1$) with the retention of a higher value of ε_t [9, 14]. Finally, in the presence of a spatial network of covalent bonds extending in all directions, which excludes blocking and ensures through transport of carriers, Eq. (8) should hold (case V; according to Eq. (9) $\beta \approx 1$).

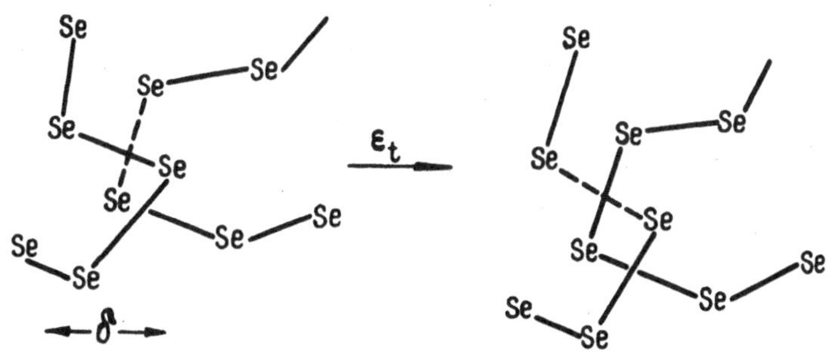

Fig. 2. Transfer of a single-electron bond from one polymeric ring to its neighbor when activated overlap of electron orbitals results from thermal atomic vibrations.

The above considerations are schematically summarized in Table 1.

In turning to a comparison of the calculated σ_0 with the experimental values σ_{0e} for various substances we should bear in mind that the theoretical value of σ_0 (8) was obtained only
with an accuracy of an order of magnitude owing to the introduction of the approximate value
$\log \nu = 13 \pm 1$. It is also significant that often a high degree of accuracy cannot be claimed for
the experimental values of σ_{0e}. It is therefore appropriate to compare the data in logarithmic
form. By Eq. (9):

$$\log \beta = \log \frac{\sigma_{0e}}{\sigma_0} = \log \frac{\sigma_{0e}}{[v]} - 4.6. \tag{10}$$

The relationships between $\log \beta$ and ε presented schematically in Table 1 provide, to a
first approximation, a satisfactory explanation of the complex conductivity variations observed
in the vitreous arsenic−selenium system. It is seen from Table 2 that an increase in the arsenic content in glasses of this system leads at first to a decrease in the number of Se−Se valence bonds and an increase in the number of As−Se valence bonds, and subsequently a decrease
of the latter leads to an increase in the number of As−As bonds. The values of $\log \sigma_{0e}/[v]$ and
$\log \beta$ are calculated taking into account the sum of the Se−Se and As−Se bonds. The values of
$\log \beta$ and ε are compared in Fig. 3.

In the region $AsSe_{12}−AsSe_{1.5}$ the energy $\varepsilon = 1.6 \pm 0.1$ eV coincides with the value $\varepsilon_i \approx$
1.6 ev found for the same compositions from the optical transmission edge (≈ 0.75 μ [10]). Therefore, in accordance with Eq. (7) the activation energy ε_a in this region must be low and not
greater than 0.1 eV. The value $\log \beta \approx -1$ for $AsSe_{1.5}$ indicates, according to Eq. (10), that σ_{0e}
is almost equal to σ_0 (within an order of magnitude). This shows the absence of any considerable cyclic formation which might block the charge carriers (case V, Table 1). This also
agrees with the x-ray data of Vaipolin and Porai-Koshits on the spatial network structure of
$AsSe_{1.5}$ [18] which allows through conduction. The decrease of $\log \beta$ to -5 as we pass from
$AsSe_{1.5}$ to $AsSe_{12}$ indicates increased blocking of the charge carriers with increasing selenium
and decreasing arsenic concentration, owing to a decrease of through conductivity. The fact
that the low value of ε_a noted above persists, indicates that conductivity is determined by the
through transport of carriers by the three-dimensional covalent network $AsSe_{1.5}$ structure which
is still present, although attenuated (case II). In the region of higher selenium content (Se−
$AsSe_{20}$) and of higher arsenic content ($AsSe_{0.6}−AsSe_{1.0}$) samples of a given composition either
have low values of ε with negative $\log \beta$, or high values of ε with high positive values of $\log \beta$.
This is a combination of cases II and III, with mutual overlapping of weakly pronounced through
conductivities with low ε_a (predominantly at low temperatures) and low conductivities in the

TABLE 1. Particular Cases of Conductivity of Polymeric Semiconductors ($\varepsilon_t \gg \varepsilon_a$, Where ε_a is Small). Effect of Structural Features of Covalently Bonded Solid

Presence of cracks and cavities (lack of continuity in solid).	Presence of covalently bonded cyclic chemical structures (ring molecules, loops, mazes, blocking of charge carriers).			Absence of cyclic covalent formations (chain or network structure with through conductivity)
	Switching of covalent bonds does not determine conductivity, which is affected by the presence of a small admixture of chain structures	Conduction determined by entropic switching of covalent bonds (critical temperature region)	Conduction determined by anentropic switching of covalent bonds (melts at elevated temperatures	
I	II	III	IV	V
$\log \beta < 0$	$\log \beta < 0$	$\log \beta > 0$	$\log \beta \approx 0$	$\log \beta \approx 0$
$\varepsilon = \varepsilon_i + 2\varepsilon_a$	$\varepsilon = \varepsilon_i + 2\varepsilon_a$	$\varepsilon = \varepsilon_i + 2\varepsilon_t$	$\varepsilon = \varepsilon_i + 2\varepsilon_t$	$\varepsilon = \varepsilon_i + 2\varepsilon_a$

TABLE 2. The Vitreous System Arsenic—Selenium [10]

Atomic composition	Density d, g/cm^3	Valence bond concentration [v] $\cdot 10^2$ (moles/cm^3)			Experimental data		Calculated data	
		[Se—Se]	[Se—As]	[As—As]	ε, eV	$\log \sigma_{0e}$	$\log \dfrac{\sigma_{0e}}{[v]}$	$\log \beta$
Se$_ж$ [15] . . .	4.3	5.4	—	—	2.3	$+3.8$	$+5.1$	$+0.5$
Se (№ 112) . .	4.29	5.4	—	—	2.0	$+4.4$	$+5.7$	$+1.1$
Se [16]	4.26	5.4	—	—	1.7	$+3$	$+4.3$	-0.3
Se [17]	4.26	5.4	—	—	1.6	$+2$	$+3.3$	-1.3
Se (№ 111) . .	4.29	5.4	—	—	1.3	-2.9	-1.6	-6.0
Se (№ 20) . .	4.29	5.4	—	—	1.0	$+0.5$	$+1.8$	-3.0
Se (№ 38) . .	4.29	5.4	—	—	0.5	-2.4	-1.1	-6.0
AsSe$_{20}$	4.33	4.81	0.78	—	1.4	-3.3	-1.2	-6.0
AsSe$_{12}$	4.34	4.41	1,26	—	1.53	-1.3	-0.1	-5.0
AsSe$_9$	4.40	4.20	1.68	—	1.53	-1.3	-0.1	-5.0
AsSe$_4$	4.44	2.85	3.42	—	1.53	-0.4	$+0.8$	-4.0
AsSe$_{2.33}$ . . .	4.54	1.55	5.24	—	1.57	$+0.5$	$+1.7$	-3.0
AsSe$_{1.5}$	4.55	—	7.08	—	1.70	$+2.2$	$+3.4$	-1.2
AsSe$_{1.25}$ . . .	4.62	—	6.66	0.66	1.5	$+1$	$+2.2$	-2.4
AsSe$_{1.0}^{*}$	4.52	—	5.88	1.47	2.2	$+5.5$	$+6.8$	$+2.2$
AsSe$_{1.0}$	4.52	—	5.88	1.47	1.3	-2.2	-1.0	-6.0
AsSe$_{0.8}^{*}$	4.55	—	5.28	2.31	2.0	$+1.6$	$+2.9$	$+1.7$
AsSe$_{0.8}$	4.55	—	5.28	2.31	1.3	-3	-1.7	-6.0

* After repeated annealing.

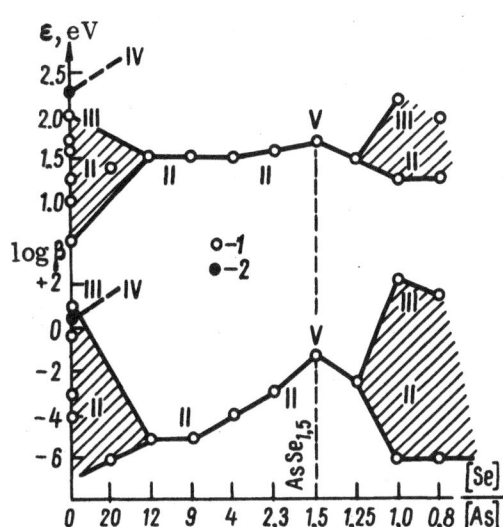

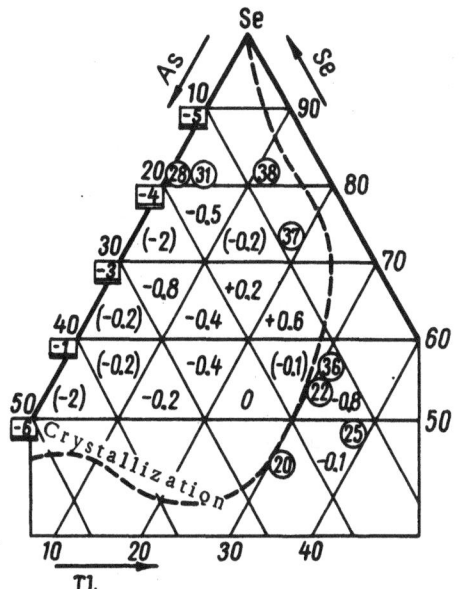

Fig. 3. Comparison of the values of ε(eV) and log β for the vitreous selenium — arsenic system. 1) Solid state (15–100°C) [10]; 2) molten selenium (250–500°C) [15]. The abscissa represents the [Se]/[As] atomic ration on an arbitrary scale.

Fig. 4. Values of log β in the conductivity of the vitreous system As—Se—Tl. The numbers in rectangles are data from Fig. 3; the numbers in circles correspond to compositions with excess of Se and TlSe (cases II and III) [21].

course of transformation of covalent bonds in cyclic structures with high ε_t with entropy effect log $\beta \approx \Delta S/4.6$ k (often at elevated temperatures). The transitions between the two conductivity types, which sometimes occur, appear in the form of kinks in the linear log $\sigma = f$ (1/T) curves. These regular effects may also be masked (especially if the temperature is raised) by shifts of equilibrium between different structures [19] and by a sharp increase of the metallic impurity conductivity as the result of spontaneous crystallization. The wide ranges of variable values of ε and log β observed in the presence of excess selenium or arsenic are attributable to a combination of these effects (Fig. 3).

Sufficiently heated molten selenium, with considerable cyclization of structure, exhibits anentropic activation of bond switching (case IV; ε = 2.3 eV, log β is close to zero; Table 2). According to the data in Table 1, the value ε_t = 0.5 (ε −1.6) $\approx$ 0.4 eV is in agreement with the viscosity-temperature relationship for selenium [20]. The approach of the experimental values of σ_{0e} to the theoretical σ_0, observed in AsSe$_{1.5}$ (owing to stabilization of the trigonal covalent network structure), is more pronounced in the more isotropic structure of vitreous AsSe$_x$Tl$_y$ of various compositions. In the vitreous region of this system the absolute values of log β do not exceed unity (Fig. 4; [21]). This is evidence of through transport of charge carriers from electrode to electrode along the covalently bonded open chains stabilized by ionized quadrupole thallium-selenide units (conductivity case V). This is consistent with the closeness between the values of the energy of electrical conduction (ε = 1.3 ± 0.1 eV) and the energy corresponding to the optical absorption edge (ε_i = 1.4 ± 0.1 eV) for the corresponding glasses [21], which indicates the low value of the activation energy $\varepsilon_a \leq 0.1$ eV according to Eq. (7).

Comparison of the experimental values of σ_{0e} for sputtered amorphous germanium and arsenic with the calculated values is worthy of attention. According to the data of Reimer [22]

TABLE 3. Conductivity of Sputtered Germanium and Arsenic after Heat Treatment (from Graphical Data of Reimer [22] and of Richter and Gommel [23])

Element	State	Concentration of valence bonds $[v] 10^2$ (mole/cm³)	ε, ev	$\log \sigma_{0e}$	$\log(\sigma_{0e}/[v])$	$\log \beta$
Ge	Amorphous, sputtered at 20°C	15	0.82 ± 0.07	2.9 ± 0.4	3.7	-0.9
	Sputtered on substrate at 400°C	15	0.76	4.1	4.9	$+0.3$
As	Amorphous, sputtered at 20°C	19	0.66	2.8 ± 0.2	3.5	-1.1
	After annealing at 100–350°C	19	0.82	2.8 ± 0.2	3.5	-1.1

and of Richter and Gommel [23] in Table 3, here again the values are in agreement to the nearest order of magnitude.

Our derivation of Eq. (8) was based on a model involving discrete localized interatomic electron clouds in the solid and took into account the known structural-chemical valence relations between atoms of different elements of the principal subgroups of the periodic table. The question of long-range order was disregarded in the derivation of Eq. (8) and therefore the possibility was not excluded that this expression might be applicable to amorphous, glassy, liquid, and crystalline semiconductors. However, the restrictions require that the above expressions relate only to intrinsic conductivity of semiconductors.

In conductivity studies on spinels of transition elements of the cobalt-manganese subgroups with and without additions of other elements, Kushnerov, Linde, and Roginskii [24] reported lack of agreement with the Verway ionic model [25]. It was noted that the valence state of the atomic-ionic chemical structural units must be the main factor determining the electrical properties of the spinels. Indeed, the stability of the electrical parameters of spinels is determined less by the constancy of the crystal structure of cubic spinels [26] than by the identity of the concentrations of covalent bonds between trivalent atoms in the octahedral cavities and oxygen atoms. The fact that the number of coordinating oxygen atoms is greater than the number of valence bonds of the trivalent atoms may be explained by statistical switching of polar covalent bonds in $(MO_{6/2})$ structural units [27] or "rotational resonance" of sd^2 bonds [28]. The experimental values of σ_{0e} for $CuMn_2O_4$ and $MnCo_2O_4$ are lower than the calculated values ($\log \beta < 0$, Table 4).

This is probably because specimens obtained by sintering of pressed tablets at high temperatures (1000–1200° [26] and 1200°C [24]) lack continuity. This explanation is consistent with the fact that for a compact of rhombohedral ferric oxide $FeO_{1.5}$ we have $\log \beta \approx 0$ (Table 4) [29].

According to Eq. (5), the mobility of the carriers has an upper limit at $\varepsilon_a = 0$ and is approximately equal to $u_0 \approx 0.1$ cm²/V·sec ($\nu \approx 10^{13}$ sec⁻¹; $\delta \approx 3 \cdot 10^{-8}$ cm; $T = 300°K$). However, at $\varepsilon_a = 0$ thermal atomic vibrations cease to influence the mobility of the charge carriers and wave-mechanical movement begins.

In discussions of activated hopping displacements of low mobility charge carriers in semiconductors some authors [2, 30] noted the analogy between such movement of electrons and holes and transport of ions in solid electrolytes.

With different initial models for ionic and electron conduction the kinetic-statistical expressions for σ_0 are identical in the two cases [31]. The numerical values are also similar.

TABLE 4. Electrical Parameters of Polycrystalline $CuMn_2O_4$, $MnCo_2O_4$, and Fe_2O_3

Atomic composition of M	Structure	$[M] \cdot 10^2$ mole/cm^3	$[v]$, mole/cm^3	Experimental data		Calculated		Literature
				ε, ev	$\log \sigma_{0e}$	$\log(\sigma_{0e}/[v])$	$\log \beta$	
$CuMn_2O_4$		2.3	0.14	0.4	+2.2	+3.0	−1.6	[26]
$CuMn_2O_4$	Cubic	2.3	0.14	0.9	+1.8	+2.7	−1.9	[24]
$MnCo_2O_4$	spinel	2.2	0.13	0.7	+3.1	+4.0	−0.6	[26]
$MnCo_2O_4$		2.2	0.13	0.9	+2.7	+3.6	−1.0	[24]
Fe_2O_3	Rhombo-hedral	3.3	0.20	2.3	+3.9	+4.6	0	[29]

The value $\log(\sigma_{0e}/[v]) = 4.6 \pm 1$ for electron semiconductors (Eq. (8)) corresponds to $\log(\sigma_0/[M]) = 3.7 \pm 1$ for ionic dielectrics, where [M] is the concentration of polar structural units in moles per cm^3 [9, 31]. The trivial difference between them is due to the introduction of the sum of the states ($Q_{he}^{1/2} = 2$) in the case of electronic conduction and to doubling of the expression in the case of bipolar intrinsic conductivity in semiconductors.

Summary

A model is proposed for the intrinsic conductivity of semiconductors with low mobility carriers, based on a concept of the discrete character of localized electron-pair chemical bonds between the atoms. The theoretical expression for the pre-exponential quantity σ_0 derived from this model gives values agreeing with the experimental results for crystalline and amorphous inorganic polymers with low mobility charge carriers. Structural features are taken into account. The common character of the mechanisms of ionic conduction in dielectrics and electron conduction in semiconductors with low mobility charge carriers is noted.

References

1. N. A. Goryunova and B. T. Kolomiets, Izv. Akad. Nauk., SSSR, ser. fiz., 20:1496 (1957); Zh. Tekhn. Fiz., 28:1922 (1958); A. R. Regel', Abstract of Doctoral Dissertation, Leningr. Gos. Univ. (1957).
2. A. F. Ioffe, Semiconductors in Modern Physics, Izd. Akad. Nauk., SSSR, Moscow−Leningrad (1954); Zh. Tekhn. Fiz., 27:1153 (1957); Fiz. Tverd. Tela, 1:157 (1959); A. F. Ioffe and A. R. Regel', Progress in Semiconductors, 4:239 (1960).
3. Semiconductors, V. P. Zhuze (editor), IL, Moscow (1960).
4. A. H. Wilson, Proc. Roy. Soc., A133:458 (1931).
5. M. Le Blanc and H. Sachse, Phys. Z., 32:887 (1931).
6. H. Krebs, Z. anorg. allg. Chem., 278:82 (1955); E. Mooser and W. B. Pearson, J. Electronics, 1:629 (1956); Canad. J. Phys., 34:1369 (1956).
7. Ya. I. Frenkel', Introduction to the Theory of Metals, Fizmatizdat, Moscow (1958).
8. H. Krebs, Acta Cryst., 9:95 (1956).
9. R. L. Myuller, Zh. Fiz. Khim., 28:1193 (1954); Zh. Prikl. Khim., 38:1077 (1955); Izv. Akad. Nauk SSSR, ser. fiz., 4:607 (1940), ● this volume, p. 170; in: The Structure of Glass, Vol. 2, Consultants Bureau, New York (1960), p. 50.
10. L. A. Baidakov, Z. U. Borisova, and R. L. Myuller, Zh. Prikl. Khim., 34:2446 (1961), ● this volume, p. 133.
11. A. R. Hippel, J. Chem. Phys., 16:372 (1948).
12. P. Rutschi, Z. phys. Chem., 14:277 (1958).
13. S. Z. Roginskii and Yu. L. Khait, Dokl. Akad. Nauk., SSSR, 130:366 (1960).
14. R. L. Myuller, Zh. Tekhn. Fiz., 25:246, 2428 (1955), ● this volume, pp. 24, 61.

15. H. W. Henkels, J. Appl. Phys., 21:725 (1950).

16. T. S. Moss, Photoconductivity in the Elements, London (1952).

17. F. Eckart, in: Halbleiterprobleme, Vol. 2, W. Schottky (editor), Braunschweig (1955), p. 69.

18. A. A. Vaipolin and E. A. Porai-Koshits, Fiz. Tverd. Tela., 2:1656 (1959).

19. H. Krebs, and F. Schultze Gebhardt, Acta Cryst., 8:412 (1955).

20. S. Dobinskii and J. Wesolowskii, Bull. Int. Acad., Polon. Sci. Lett., 10–11A:449 (1936).

21. R. L. Myuller and T. P. Markova, Vestn. Leningr. Gos. Univ., Vol. 17, No. 4, Iss. 1, p.75 (1962).

22. L. Reimer, Z. Naturforsch., 13a:536 (1958).

23. H. Richter and G. Commel, Z. Naturforsch., 12a:996 (1957).

24. M. Ya. Kushnerov, V. R. Linde, and S. Z. Roginskii, Fiz. Tverd. Tela, 3:384 (1961).

25. E. J. W. Vervey, P. W. Haaijam, F. S. Romeijn, and G. W. van Oosterhout, Philips Res. Rep., 5:173 (1950).

26. I. T. Sheftel', A. I. Zaslavskii, E. V. Kurlina, and T. I. Tekster-Proskuryakova, Fiz. Tverd. Tela. 1:227 (1959).

27. R. L. Myuller, Zh. Tekhn. Fiz., 25:9, 1556 (1955), ● this volume, p. 33; Izv. Tomsk. Politekhn. Inst., 91:239 (1956).

28. S. L. Altmann, C. A. Coulson, and W. Hume-Rothery, Proc. Royal Soc., A240:145 (1957); T. A. Kontorova, Fiz. Tverd. Tela, 1:1761 (1959).

29. F. J. Morin, Bell System Tech. J., 37, 1047 (1958).

30. A. R. von Hippel, Ergebn. Ex. Naturwiss., 14:79 (1935); Z. Phys., 101:680 (1936); Dielectrics and Waves, New York–London (1954); R. R. Heikes and W. D. Johnston, J. Chem. Phys., 26:582 (1957); J. Phys. Chem. Solids, 7:1 (1958).

31. Ya. I. Frenkel', Z. Phys., 35:652 (1926); W. Jost, J. Chem. Phys., 1:466 (1933); R. L. Myuller, Zh. Fiz. Khim., 6:616 (1935); Fiz. Tverd. Tela, 2:1333 (1960); 2:103 (1935); The Structure of Glass, Vol. 2, Consultants Bureau, New York, (1960), p. 215.

Electrical Conductivity in the Vitreous Selenium-Arsenic System

Difficulties in the physical interpretation of the electron conductivity of amorphous materials prompted A. F. Ioffe and other workers to put forward short-range order and the associated chemical bonding as the most important factors determining the nature of semiconductors [1-7]. Investigations of liquid [8-10] and vitreous [11-13] semiconductors have played a considerable part in studies of the semiconductor problem in this direction. At the same time, studies of the problem of the vitreous state of matter also revealed the cardinal role of the nature of the chemical bond in glass formation [14].

The growing interest in the chemical factors which determine the physical properties of semiconductors makes it necessary to expand considerably the physicochemical experimental investigations. Together with problems of the influence of all types of disturbances on an ordered and chemically homogeneous crystal lattice (dislocations, impurities, deviations from stoichiometry, etc.) new problems arise concerned with essentially chemical structural changes taking place in solids during heat treatment, irradiation, etc. For example, problems associated with the degree of lattice order (in single crystals and polycrystalline bodies) have always attracted the attention of physicists. At the same time the chemical aspects of the transition of an amorphous into a crystalline material remained obscure, although there is no doubt that ordering of a structure is closely associated with the breakdown of chemical bonds between atoms and the formation of new bonds. There is reason to believe that such processes do not involve the complete rupture of interatomic bonds, but occur by bond switching which requires a lower activation energy [15, 16].

This view throws a new light on a number of facts including the well known difficulty of obtaining single crystals of such a simple elementary substance as selenium. Polymeric transformations involving switching of valence bonds (isomerization) which undoubtedly limit crystallization processes, must also be of some significance in relation to the known fact that it is impossible to obtain systematically reproducible conductivity data for selenium samples from different melts and with different thermal histories [17].

It was of interest to study the possibility of stabilizing the conductivity of selenium by cross-linking its chains by atoms of multivalent elements. Arsenic and germanium are especially suitable among such elements, as they are immediate neighbors of selenium in the Periodic Table and thus ensure that the chemical bonds between atoms of these elements are almost entirely homopolar.

There is reason to believe that quenched glassy selenium contains Se_8 ring molecules, Se_n cyclic structures, and a small percentage of open $-Se_n-$ spiral chains which partially ensure through conduction. At the same time, some of these open chains may form complex "mazes" with chains terminating within the glass, far from the surface, and thus create cul de sacs for the current carriers. These cul-de-sacs, like cyclic structures, block the current carriers, and may give rise to polarization effects [18]. This complex structure in selenium is

*L. A. Baidakov, Z. U. Borisova, and R. L. Myuller, Zh. Prikl. Khim., 34:2446 (1961).

133

highly sensitive to a heat treatment which favors ordering of the structure with the conversion of the blocking cyclic structures and mazes into the crystalline hexagonal modification with through conduction along the chains.

Arsenic can have a two-fold effect in selenium. It may, in the course of network formation, close the chains into complex polycycles of the type As_4Se_{6n} and thus favor blocking of the current carriers. Arsenic can also give rise to an open $AsSe_{3n}$ network which ensures through conduction. In both cases a network is formed, with a sharp increase of viscosity (Krebs [17]). It is also evident that with the transformation to the composition $AsSe_{1.5}$ the blocking of the current carriers by cyclic selenium should cease and the network structure ($AsSe_{3/2}$) become open for the current carriers.

Vitreous selenium–arsenic alloys [19], in the As_xSe composition range where $0 \le x \le 1.25$, were prepared from arsenic and selenium (both redistilled twice under vacuum) by vacuum fusion in ampoules using the method of Goryunova and Kolomiets [11]. The ampoules were heated at a rate of 170-180° per hour, held at 680-700° with vibration [20] for 2 hours, and then cooled at the rate of 0.5° per minute. In some duplicate preparations the specimens were held at 680-700° for 4-5 hours without vibration, but were additionally annealed for 2 hours at 280°. This melting procedure for glasses Nos. 2-8 of composition $AsSe_y$, where $1.25 \le y \le 20$ gave specimens with satisfactorily reproducible conductivity-temperature relationships and with similar values of the energy ε and the pre-exponential term σ_0 in the conductivity expression $\sigma = \sigma_0 \exp(\varepsilon/2kT)$.

The density d and microhardness H provide macroscopic characterization of the vitreous alloys so obtained. These values are given in Table 1 and are close to the values for glasses in the same system synthesized by Aio and Kokorina [21] for another purpose. The data in the table can be used to calculate the mole-volume values of the microhardness for the structural elements

$$SeSe_{2/2}(-Se-Se-Se-),$$

$$AsSe_{3/2}\left(-Se-As\genfrac{}{}{0pt}{}{Se-}{Se-}\right) \text{ and } As_2Se_{4/2}\left(\genfrac{}{}{0pt}{}{-Se}{-Se}As_2\genfrac{}{}{0pt}{}{Se-}{Se-}\right),$$

$$h_{Se_2} = \frac{H_{Se_2}}{[Se_2]} = 1.56 \cdot 10^3 \,\text{kg/mm}^2 \cdot \text{cm}^3/\text{mole} \,(\pm 3\%),$$

$$h_{AsSe_{1.5}} = \frac{H_{AsSe_{1.5}}}{[AsSe_{1.5}]} = 5.9 \cdot 10^3 \,\text{kg/mm}^2 \cdot \text{cm}^3/\text{mole} \,(\pm 6\%)$$

and

$$h_{As_2Se_2} = \frac{H_{As_2Se_2}}{[As_2Se_2]} = 6\,6 \cdot 10^3 \,\text{kg/mm}^2 \cdot \text{cm}^3/\text{mole} \,(\pm 8\%),$$

where

$$[Se_2] = \frac{d}{157.9} \,\text{mole/cm}^3; \quad [AsSe_{1.5}] = \frac{d}{193.4} \,\text{mole/cm}^3;$$

$$[As_2Se_2] = \frac{d}{307.7} \,\text{mole/cm}^3.$$

Comparison of the experimental and calculated values shows that the microhardness of glasses Nos. 1-7 in the $Se-AsSe_{1.5}$ composition range obeys the additivity law:

$$H\,\text{cal.} = h_{Se_2}[Se_2] + h_{AsSe_{1.5}}[AsSe_{1.5}] \,\text{kg/mm}^2,$$

where

$$[AsSe_{1.5}] = [As]$$

and

$$[Se_2] = 0.5([Se] - 1.5[As]).$$

TABLE 1. The Vitreous Selenium—Arsenic System

Glass No.	Atomic composition	A content (atomic %)	Content in (mole/ml) .100			Density d (g/cm²)	Microhardness H kg/mm²		Electrical conductivity		Optical absorption edge, μ	Remarks (number of specimens studied in parentheses)
			[Se]	[As]	Total		Exptl.	Calc.	ε, ϑϵ	log σ₀		
1	Se	—	5.43	—	5.43	4.29	42±1	—	0.97	+0.5	—	Glass from melt No. 20
									0.45	−2.4	—	Glass from melt No. 38
									1.3	−2.9	0.70	Glass from melt No. 111, specially pure, quenched
1	Se	—	—	—	—	4.26	—	—	2.04	+4.4	—	Glass from melt No. 112, specially pure, quenched
						—			1.7	+3	—	
									1.6	+2	—	(Moss (Barnard))
2	AsSe$_{30}$	4.8	5.20	0.26	5.46	4.33_1	52±3	53	1.4	−3.3	0.74	(2)
3	AsSe$_{12}$	7.7	5.04	0.42	5.46	4.34_2	59±3	59			0.75	(3)
4	AsSe$_9$	10.0	5.04	0.56	5.60	4.40±0.04	65±2	66	1.53±0.02	−1.3∓0.1	0.78	(2)
5	AsSe$_4$	20.0	4.56	1.14	5.70	4.44_8	87±5	90	1.53±0.02	−0.4±0.4	0.73	(2)
6	AsSe$_{2.33}$	30.0	4.17	1.75	5.92	4.54±0.01	112±4	116	1.57±0.02	+0.5±0.3	0.73	(4)
7	AsSe$_{1.5}$	40.0	3.54	2.36	5.90	4.55±0.03	140±9	—	1.70±0.03	+2.2±0.4	0.74	(3)
8	AsSe$_{1.25}$	44.4	3.33	2.66	5.99	4.62_0	127±11	123	1.5±0.1	+1±1	—	(2)
9	AsSe$_{1.0}$	50.0	2.94	2.94	5.88	4.52_2	97±8	—	1.3	−2.2	0.74	Repeated annealing (1)
	AsSe$_{1.0}$	50.0	2.94	2.94	5.88	4.52_2			2.18	+5.5		
10	AsSe$_{0.8}$	55.6	2.64	3.30	5.94	4.55±0.03	66±5	—	1.25	−3.0	—	Repeated annealing (3)
	AsSe$_{0.8}$	55.6	2.64	3.30	5.94	4.55±0.03			1.95±0.04	+1.6±0.1	—	

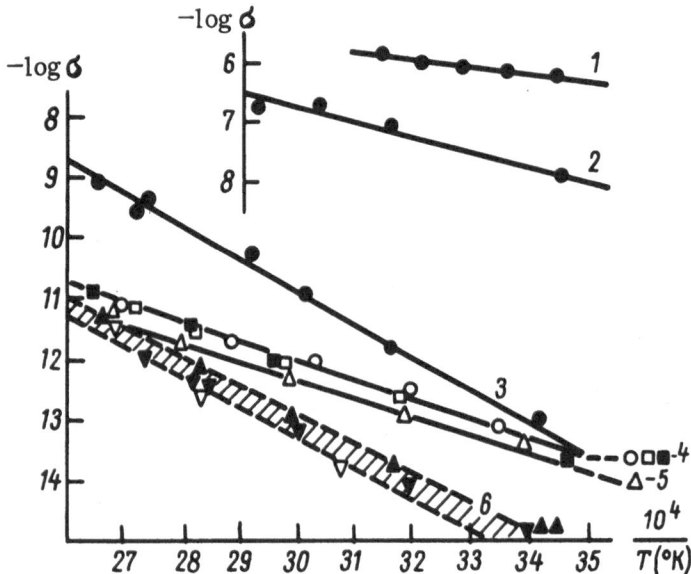

Fig. 1. Temperature dependence of conductivity:
1) Se No. 38; 2) Se No. 20, glasses after repeated annealing at 180°; 3) AsSe; 4) $AsSe_{0.8}$; without repeated
annealing: 5) AsSe, 6) $AsSe_{0.8}$.

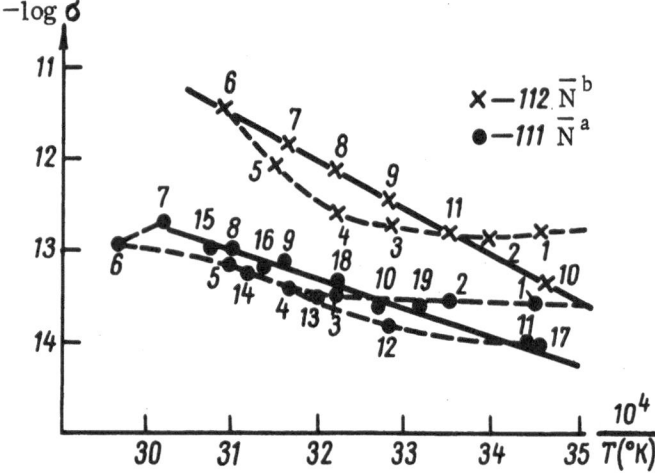

Fig. 2. Temperature dependence of conductivity for
quenched specimens of high purity vitreous selenium
(the numbers indicate the measurement sequence;
consecutive measurements were made at intervals of
about one hour). Glass melts: a) 111; b) 112.

The maximum microhardness of $AsSe_{1.5}$ indicates the existence here of a network-poly-
mer compound. The monotonic variation of microhardness indicates the absence of compounds
of the $Se^-As^+Se_{3/2}$ (or As_2Se_5) type. The higher solution rate of $AsSe_{2.33}$ which was noted earlier is evidently the consequence of the loosening of the $AsSe_{1.5}$ network structure by one-dimen-
sional chain inclusions of excess selenium [22]. The microhardness of glass No. 8 ($AsSe_{1.25}$)

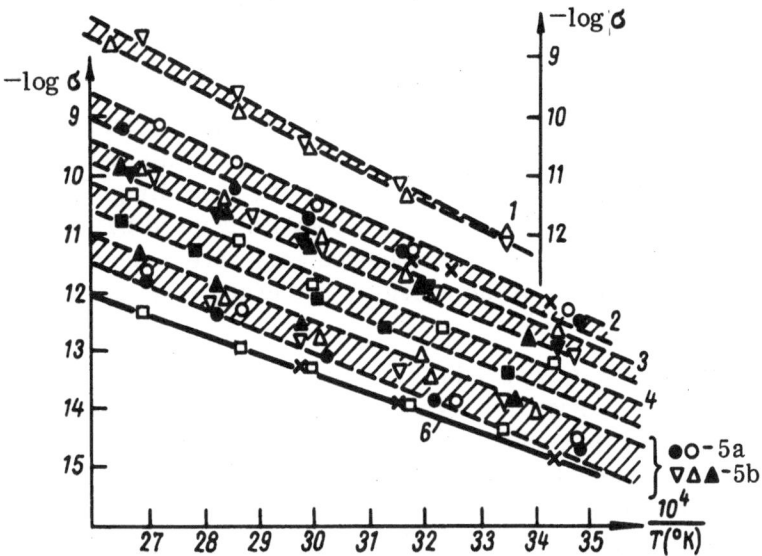

Fig. 3. Temperature dependence of conductivity of vitreous arsenic selenides. 1) $AsSe_{1.25}$ (2 specimens); 2) $AsSe_{1.5}$ (3 specimens); 3) $AsSe_{2.33}$ (4 Specimens); 4) $AsSe_{4.0}$ (2 specimens); 5a) $AsSe_{9.0}$ (2 specimens); 5b) $AsSe_{12.0}$ (3 specimens); 6) $AsSe_{20}$ (2 specimens). Data for different melts of the same compositions are denoted by different symbols.

satisfies the additive relationship:

$$H_{cal} = h_{AsSe_{1.5}} [AsSe_{1.5}] + h_{As_2Se_2} [As_2Se_2] \text{ kg/mm}^2.$$

The decrease of the microhardness on the transition from $AsSe_{1.5}$ to As_2Se_2 is due to the appearance of relatively weak As—As bonds in the $As_2Se_{2/2}$ structural units. At the same time the viscosity of As_2Se_2 should be higher because of the increase of the activation energy of switching of chemical valence bonds in the transition from trigonal $AsSe_{3/2}$ to tetrahedral structural units. The microhardness of $AsSe_{0.8}$ is apparently determined entirely by the $As_2Se_{4/2}$ structural units.

The low conductivities of the glasses in the system under investigation made it necessary to use an electrometric method for conductivity measurements. With specimens having a resistance greater than 10^{11} ohms the charge–discharge of a small capacitor was used, with allowance for the insulation leakage of 10^{13}-10^{14} ohms of the electrometric part of the system. The resistance was calculated from the formula

$$R_x = \frac{V_p R_z}{V} \left(1 - e^{-\frac{t}{CR_z}}\right),$$

where V_p is the potential applied to the specimen, V the potential of the capacitor C at time t, and R_z the insulation resistance [23].

The electrodes were brass disks from 0.80 to 1.30 cm^2 in area at the base. Silver paste was used to contact the glass specimens. Control experiments with Aquadag paste ($AsSe_4$ glass) showed that, within the limits usual for this system, the reproducibility of the temperature—conductivity relationship was satisfactory.

The electroded specimens were put in a tube furnace and the apparatus did not differ essentially from that used earlier by one of the present authors [23]. The temperature was automatically controlled and the fluctuations did not exceed ±0.5°.

The electrical conductivity of the glass specimens studied are given in Figs. 1-3. Determinations were carried out for each specimen both by comparison with standard resistances and by capacitor charge and discharge. Measurements with rising and falling temperatures were made in order to confirm the absence of hysteresis. About 1 h elapsed between consecutive measurements at different temperatures. Parallel measurements were made with two to four specimens from different melts. The values of the energy ε and of log σ_0 derived from Figs. 1-3 are given in the table.

It was noted earlier that conductivity data for $AsSe_x$ glasses with $1.25 \leq x \leq 20$ are satisfactorily reproducible. The effect of impurities on the conductivity of these glasses was investigated by means of parallel melts of $AsSe_{1.5}$ and $AsSe_{2.33}$ glasses with highly purified V-4 grade arsenic and selenium. The determinations showed that the conductivity–temperature relationships of the glasses did not differ substantially from those for the same glasses synthesized from double-distilled arsenic and selenium. Kolomiets and Nazarova [24] found earlier that impurities do not have any appreciable influence on the conductivity of other chalcogenide glasses.

Selenium which had been double-distilled under vacuum (about 0.1 mm) and annealed after melting had an electrical conductivity of about 10^{-6}-10^{-8} Ω^{-1} cm^{-1} at 20°, with ε ≈ 0.5-1.0 eV (Fig. 1, selenium from melts Nos. 20 and 38). The conductivity must have been high because of increased through conduction along valence-bonded atomic chains as the result of a partial crystallization during annealing resulting from the catalytic influence of oxygen and other impurities [1, 25]. Specimens of vitreous selenium from melts Nos. 111 and 112, made from V-4 material with rapid cooling after melting under vacuum at 10^{-3} - 10^{-4} mm, had $\sigma \approx 10^{-13}$ - 10^{-14} $\Omega^{-1} \cdot$ cm^{-1} and ε ≈ 1.3−2.4 eV at 20°. These values, which are in satisfactory agreement with the results reported by others [17, 18, 26, 27], were obtained below 60-70°, the temperatures at which softening and crystallization begins [17, 28, 29].

Figure 2 gives a general picture of the results obtained for selenium glasses of melts Nos. 111 and 112 at various temperatures. Since the interval between two consecutive measurements was about one hour it may be assumed that the delayed Kozyrev effect [17] was not appreciable. Figure 2 shows that the initial measurements involved complex hysteresis effects. These effects disappeared after 6-15 hours of moderate heating at 40-60°. In the complex combination of effects we must note the initial higher conductivity at room temperature, most likely due to surface moisture [30], lost after moderate heating. Selenium is acted on by oxygen and moisture at room temperature when exposed to light in humid air, with the formation of SeO_2 [31], and very probably the following thermodynamic equilibrium is established [32]:

$$H_2SeO_3 \rightleftarrows H_2O + SeO_2 - 0.9 \text{ kcal.}$$

After removal of the moisture the conductivity falls if the surface SeO_2 content is fairly low.

Satisfactorily reproducible conductivity–temperature relationships could not be obtained for vitreous AsSe and $AsSe_{0.8}$. The data in Table 2 and Fig. 1 show that a second anneal for 2-5 h at 180° resulted in an increase of the temperature coefficient of conductivity. Subsequent annealings did not influence the conductivity. It is noteworthy that normally annealed AsSe and $AsSe_{0.8}$ glasses have ε ≈ 1.2 - 1.3 before the second annealing, which corresponds to the value for amorphous arsenic [27, 33].

$AsSe_x$ glasses with $1.25 \leq x \leq 20$ did not exhibit hysteresis in the conductivity curves, which were not influenced by repeated annealing. It must be emphasized that the conductivity of these glasses remained relatively stable despite the fact that the range of their σ values as a function of $1/T$ (Fig. 3) lies entirely in the wide range of varying values of the conductivity of selenium. The data for the arsenic−selenium glasses are somewhat scattered, but the variability is incomparably less than the enormous variations of the conductivity of vitreous selenium (Figs. 1 and 2). It is therefore evident that arsenic has a pronounced stabilizing effect on conductivity. This is readily explained by the formation of the three−dimensional covalently bonded network polymer structure of arsenic selenide, which allows through transport of the current carriers.

The value of the energy $\varepsilon = 1.70 \pm 0.03$ eV found for the glass $AsSe_{1.5}$, falls for glasses of closely related composition $AsSe_{1.25}$ and $AsSe_x$ ($2.3 \leq x \leq 20$) to 1.5 ± 0.1 eV.

Table 1 shows that the absorption edge for all the glasses studied is about 0.75 ± 0.02 μ which corresponds to 1.65 ± 0.05 eV, in close agreement with the temperature dependence of conductivity.

The values of $\varepsilon = 1.70$ eV and $\log \sigma_0 = 2.2 \pm 0.4$ differ from the corresponding values of 2.0 eV and 3.7 derived from the graphical data of Goryunov and Kolomiets [34].

The correlation between variations of the energy ε and the exponential statistical factor [35] has already been noted in relation to the ionic conductivity and viscosity of glasses [36]. It was then attributed to the entropy of activation of the respective processes. Even earlier, Borelius and his co-workers [26] established such a correlation for the conductivity−temperature relationship of selenium after heat treatment and under the influence of impurities which catalyze structural changes and crystallization. Transformation of the Borelius function gives the expression $\varepsilon = 1.04 + 0.26 \log \sigma_0$. The results of the present investigation are similarly correlated in Fig. 4. This gives the expression $\varepsilon = 1.60 + 0.12 \log \sigma_0 \pm 0.16$, which differs from that found by Borelius for selenium. The points for selenium (glass melts Nos. 20, 38, and 112) found in the present investigation are close to the straight line obtained by Borelius. One point, for glass melt No. 111, lies in the region corresponding to glasses with network structures. These important relationships require a special kinetic investigation.

The results of the present investigation of electrical conductivity, reviewed in the light of the structural and valence changes taking place in the Se−As glassy system with an increase of the [As]/[Se] atomic ratio from 0 to 1.25, can be summarized as follows:

The electrical conductivity of selenium with a low impurity content is unstable, and varies over a range of ten orders of magnitude. This variability of the electrical conductivity is due to the ease with which structural transformations

$$n\,(Se_8) \rightleftarrows \frac{8n}{n_i}(Se_{n_i}) \rightleftarrows \ldots \rightleftarrows \frac{8n}{n_k}(-Se_{n_k}-) \rightleftarrows \frac{8n}{n_k}(-Se-)_{n_k}$$

occur as the result of shifts in the complex multistage equilibrium with the reorganization of Se_8 rings and of a whole spectrum of n_i-membered cyclic polymers characteristic of liquid selenium, into multimembered chain structures, at first disordered, and then ordered and characteristic of crystalline hexagonal selenium. The total number of covalent bonds remains essentially unchanged in the process, so that the shift of equilibrium involves only a slight energy effect.

The sharp decrease of conductivity on melting hexagonal selenium [26] and the analogous fall of conductivity on the transition from hexagonal selenium ($-Se_n-$, $\sigma = 10^{-5} - 10^{-7}$ $\Omega^{-1} \cdot cm^{-1}$ [37]) to monoclinic selenium (Se_8, $\sigma \approx 10^{-15} \Omega^{-1} \cdot cm^{-1}$ [38]) show that the enormous range of the electrical conductivity of selenium, from 10^{-15} to $10^{-5} \Omega^{-1} \cdot cm^{-1}$, is primarily the result of conversion of ring structures, which block the current carriers, into open chains which ensure through transport of vacancies and electrons.

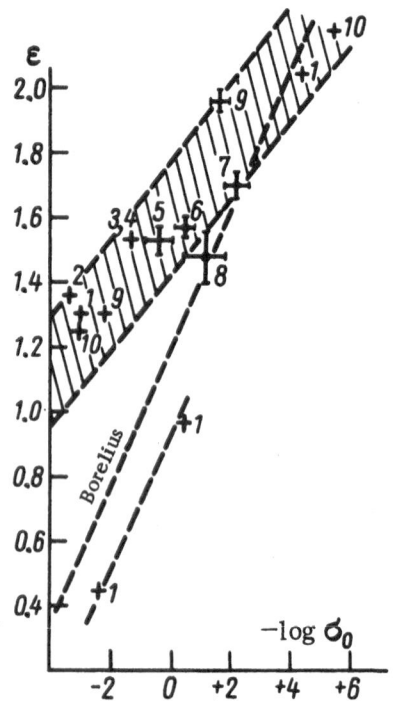

Fig. 4. Correlations between ε and σ_0. The numbers correspond to the numbers of the glasses (see Table 1).

The highly complex intermediate structural and valence states of quenched vitreous selenium are obtained at various stages of annealing as the result of activated switching of covalent bonds when the electron clouds of the bonds overlap as a result of torsional vibrations [14]. The appropriate heat treatment gives rise either to the molecular Se_8 rings and multiatomic Se_n cyclic molecules of variable composition (the average value of n = 100 [39]) characteristic of liquid selenium, or to the hexagonal chain structures characteristic of crystalline selenium.

Conductivity data indicate that the introduction of arsenic "fixes" the network structure, which is considerably more stable, has a greater hardness, a higher softening temperature [21], and a more stable electrical conductivity. The vitreous compound $AsSe_{1.5}$, with the maximum density of spatial packing of covalent bonds, has the optimum strength and most stable conductivity. A further increase in the arsenic content reduces the microhardness and the conductivity again becomes unstable owing to breakdown of the relatively stable $AsSe_{1.5}$ structure.

The blocking of the current carriers in ring and cyclic structures should influence the magnetic properties of the glasses in question. In particular, it is likely that the considerable difference between the atomic magnetic susceptibility of selenium ($-26.5 \cdot 10^{-6}$) and arsenic ($-5.5 \cdot 10^{-6}$) [40] is associated, to a considerable extent, with the structural and valence features of these solid elements.

An investigation of the influence of the thermal history on the magnetic properties of the vitreous arsenic—selenium system is of exceptional interest.

Summary

1. It has been experimentally established that introduction of arsenic into selenium stabilizes the electrical conductivity because of conversion of a labile valence-chain structure in selenium into a stronger and more stable valence-network structure.

2. Glasses of the arsenic—selenium system exhibit correlation between the energy ε and the statistical parameter log σ_0, in accordance with the equation

$$\varepsilon = 1.60 + 0.12 \log \sigma_0 \pm 0.16.$$

3. Glasses of the compositions AsSe and $AsSe_{0.8}$ are labile and measurements of their conductivity are not satisfactorily reproducible.

4. The initial conductivity measurements on quenched selenium at temperatures below 60° exhibit a hysteresis effect, caused by the influence of moisture and eliminated by slight heating.

References

1.   A. F. Ioffe, Anthology Dedicated to the 30th Anniversary of the Great October Socialist Revolution, Izd. Akad. Nauk. SSSR, 1947, p. 305.

2. A. F. Ioffe, Zh. Tekhn. Fiz., 27:1153 (1957); Fiz. Tverd. Tela, 1:157 (1959).
3. A. F. Joffe (Ioffe) and A. R. Regel (Regel'), Progress in Semiconductors, 4:239 (1960).
4. H. Krebs, Acta Cryst., 9:95 (1956).
5. E. Mooser and W. B. Pearson, J. Electronics, 1:629 (1956).
6. C. H. L. Goodman, Phys. Chem. Solids, 6:305 (1958).
7. W. P. Zhuse (editor), in: Semiconducting Materials (Problems of the Chemical Bond), IL, Moscow (1960).
8. A. R. Regel', in: Theoretical Problems and Studies in Semiconductor Metallurgy, Izd. Akad. Nauk. SSSR, Moscow (1955), p. 12; A. R. Regel', Doctoral Dissertation, LGU (1957).
9. A. I. Blum and A. R. Regel', Zh. Tekhn. Fiz., 21:316 (1951); 23:783, 964 (1953).
10. A. I. Blum, N. P. Mokrovskii, and A. R. Regel', Izv. Akad. Nauk. SSSR, ser. fiz.,16:139 (1952).
11. N. A. Goryunova and B. T. Kolomiets, Izv. Akad. Nauk. SSSR, ser. Fiz., 20:1496 (1956); Zh. Tekhn. Fiz., 28:1922 (1958).
12. T. I. Vengel' and B. T. Kolomiets, Zh. Tekhn. Fiz., 27: 2484 (1957).
13. N. A. Goryunova, B. T. Kolomiets, and V. P. Shilo, Zh. Tekhn. Fiz., 28:981 (1958).
14. R. L. Myuller, Izv. Akad. Nauk, SSSR, ser. fiz., 4:607 (1940), ● this volume, p. 170; in: The Structure of Glass, Vol. 2, Consultants Bureau, New York (1960), p. 50.
15. R. L. Myuller, Zh. Prikl. Khim., 28:1077 (1955).
16. V. T. Slavyanskii, in: The Structure of Glass, Vol. 2, Consultants Bureau, N. Y. (1960), p. 289.
17. B. Gudden, Ergeb. exakt. Naturwiss., 3:116 (1924); H. Krebs, Z. anorg. allg. Chem., 265: 156 (1951); R. Yu. Koden, Zh. Éksp. Teor. Fiz., 24:714 (1953); F. Eckart, Ann. Phys., 14:233 (1954); 17:84 (1956); in: Halbleiterprobleme, Vol. 2, W. Schottky (editor), Braunschweig (1955), p. 69; F. Brunke, ibid., p. 59; P. T. Kozyrev, Zh. Tekhn. Fiz., 26:255 (1956); Abstract of Doctoral Dissertation, Leningrad, 1958; T. Sekiguti, Reports of Sci. Res. Inst. Japan. Phys. Chem. Sect.,33:115 (1957); Sci. Papers Inst. Phys. Chem. Res. Tokyo, 53:120 (1959); T. S. Moss, Photoconductivity in the Elements, London, (1952).
18. D. Nasledov and E. Malyshev, Zh. Tekhn. Fiz., 16:1127 (1946).
19. S. S. Flaschen, A. D. Pearson, W. R. Northovers, J. Am. Cer. Soc., 42:450 (1959).
20. A. S. Borshchevskii and D. N. Tret'yakov, Fiz. Tverd. Tela, 1:1483 (1959).
21. L. G. Aido and V. F. Kokorina, Optiko-mekhanic. Prom., 28(4):39 (1961).
22. R. L. Myuller, Z. U. Borisova, and N. I. Grebenshchikova, Zh. Prikl. Khim, 34:533 (1961).
23. S. A. Shchukarev and R. L. Myuller, Zh. Fiz. Khim, 1:625 (1930); Z. phys. Chem., A150: 439 (1930).
24. B. T. Kolomiets and T. F. Nazarova, in: Physics of Solids, Vol. 2, Izd. Akad. Nauk, SSSR, Moscow—Leningrad (1959), p. 22; Fiz. Tverd. Tela, 2:174 (1960).
25. G. Blet and D. Vidal, C. R. Acad. Sci., 249:2068 (1959).
26. G. Borelius, F. Pihlstrand, J. Anderson, and K. Gullberg, Ark. Mat. Astron. Fys., 30A (14):1 (1944).
27. N. N. Sirota, in: Pure Metals and Semiconductors, Metallurgizdat, Moscow (1959), p. 22.
28. G. Tammann, R. Klein, Z. anorg. allg. Chem., 192:171 (1930); Z. Elektrochem, 36:674 (1930).
29. K. Tanaka, Mem. Sci. Kyoto Univ., A17:77 (1934); K. Tanaka and H. J. Tien, ibid., A18: 309 (1935).
30. A. O. Rankine and J. W. Avery, Nature, 118:13 (1926).
31. G. Caleagni, Gazz. Chim. Ital., 65:558 (1935); Gmelins Handbuch der anorganischen Chemie, 8 Auflage, Syst. No. 10 Teil A, Lief. 1 (1952), p. 246.
32. D. Rossini, Selected Values of Chemical Thermodynamic Properties, Nat. Bur. Stand. Circ. No. 500 (1952).
33. A. Becker and I. Schaper, Z. Phys., 122:49 (1944).

34. N. A. Goryunova and B. T. Kolomiets, Zh. Tekhn. Fiz., 25:2070 (1955).

35. P. Rutschi, Z. phys. Chem., Frankfurt, 14:277 (1958).

36. R. L. Myuller, Zh. Tekhn. Fiz., 25:2428 (1955), ● this volume, p. 61; Zh. Prikl. Khim., 28:1077 (1955); Fiz. Tverd. Tela, 1:346 (1959); 2:1345 (1960), ● this volume, pp. 105, 112; in: Physics of Dielectrics, Izd. Akad Nauk, SSSR, Moscow (1960), p. 445.

37. K. W. Plessner, Proc. Phys. Soc. B64:671 (1954).

38. B. Gudden and R. W. Pohl, Z. phys. Chem. 48:384 (1928).

39. H. Krebs., Z. Phys., 126:768 (1949); H. Krebs and W. Morsch, Z. anorg. allg. Chem., 263: 305 (1950).

40. P. W. Selwood, Magnetochemistry, Interscience, New York (1956).

Electrical Conductivity in the Vitreous As-Se-Ge System*

1. The electron conductivity of a solid is essentially determined by its chemical nature; that is, primarily by its composition and the nature of the chemical bonds between its atoms. In order to resolve the problems of solid state electrochemistry it is essential to study vitreous solids. It is known that the vitreous state is closely associated with the presence of mainly short range valence bonds and the absence of an ordered crystalline structure in the solid. The latter helps to remove the effects of mechanical defects which are statistically difficult to control. The disordered three-dimensional network structure of covalently bonded atoms in a stable vitreous solid is particularly amenable to the application of the principles of statistical physics. It is very interesting, from this point of view, to study the conductivity of the Ge—As—Se vitreous system. As these elements belong to the principal IV to VI subgroups and occupy the neighboring positions 32-33-34 in the Periodic Table, the atomic interactions are mainly covalent and produce an extensive glass-forming region.

The previous studies of the two-component vitreous semiconducting systems As—Se [1], Ge—Se [2], and As—Se—Tl [3], indicated that on the transition from the covalent chain structures of selenium to the network structure of the selenides, the value of the conductivity modulus $\log(\sigma_{0e}/[v])$ approached 4.6 ± 1 expected from the valence nature of the conductivity of vitreous semiconductors [4]. Here $[v]$ is the concentration of covalent bonds in mole/cm^3, and σ_{0e} is the experimentally determined frequency factor in the expression for the temperature dependence of specific conductivity

$$\sigma = \sigma_{0e} \exp\left(-\frac{\varepsilon_\sigma}{2kT}\right), \tag{1}$$

where $\varepsilon_\sigma = \varepsilon_i + 2\varepsilon_a$ (ε_i is the ionization energy of the individual covalent bond; and ε_a is the energy of the small potential barrier to the displacement of low mobility carriers).

The present paper deals with Se—As—Ge glasses which do not contain excess selenium. In common with the systems studied previously, As—Se, Ge—Se, and As—Se—Tl, the study of the As—Se—Ge system has not been limited to compositions which correspond to the pseudo-binary division of the diagram but has been extended into an appreciable region of the glass forming compositions range.

As a result of the mutual saturation of the valency of the atoms, the continuous network structure typical of a vitreous solid forms during the melting of the initial materials, [5]. In this As—Se—Ge system, the difference in the spatial structure of the trigonal As—Se— and the tetrahedral Ge structural units (s.u.) leads to the destruction of the valence coordination periodicity in the solid which must serve as a steric disturbance to the formation of the

---

*R. L. Myuller, L. A. Baidakov, and Z. U. Borisova, Vestn. Leningr. Gos. Univ., No.10, p.94 (1962).

TABLE 1. Data for the Vitreous As-Se-Ge System

Class No.	At % (batch) As	Se	Ge	Density d, g/cm³	Chemical composition (m = AsSe$_x$Ge$_y$) x	y	Mole wt. M	Concentration [M] 10², mole/cm³	GeSe$_{4/4}$	AsSe$_{3/2}$	AsSe$_{3/2}$	AsAs$_{3/3}$	GeGe$_{4/4}$	Microhardness H, kg/mm² Experimental	Calculated	Valence bonds Ge-Se	As-Se	Energy ε$_\gamma$	ε$_\sigma$	log σ$_{0e}$	log(σ$_{0e}$/[γ])	log β
1	35.7	60.7	3.6	4.594	1.70	0.10	216	2.12	0.21	2.12	—	—	—	149±5	152	(0.84)	6.36	1.7	1.8	+3.2	+4.4	-0.2
2	32.2	61.3	6.5	4.56₆	1.90	0.20	240	1.91	0.38	1.91	—	—	—	154±3	160	(1.52)	5.73	1.7	1.8	+1.8	+3.0	-1.6
3	25.0	62.5	12.5	4.50	2.50	0.50	309	1.46	0.73	1.46	—	—	—	154±9	176	(2.92)	4.38	1.7	1.7	-0.2	+1.2	-3.4
4	18.2	63.6	18.2	4.44	3.50	1.00	424	1.05	1.05	1.05	—	—	—	196±8	189	(4.20)	3.15	1.9	1.7	-1.1	+0.4	-4.2
5	14.3	64.3	21.4	4.36₅	4.50	1.50	539	0.81	1.22	0.81	—	—	—	181±4	194	4.88	(2.43)	1.9	2.0	+1.1	+2.4	-2.2
6	19.3	58.0	22.7	4.39₅	3.00	1.17	397	1.11	1.30	—	0.73	0.19	0.35	208±4	209	5.20	—	—	2.0	—	+2.4	-2.2
7	27.2	40.9	31.9	4.58₃	1.5	1.17	278	1.65	1.24	—	0.25	0.82	—	273±9	272	4.96	—	1.8	2.2	+4.5	+5.8	+1.2
8	23.9	52.2	23.9	4.42₉	2.18	1.00	320	1.38	1.38	—	—	0.56	0.48	241±11	229	5.52	—	1.8	2.2	+2.7	+4.0	-0.6
9	19.7	42.9	37.4	4.58₄	1.18	1.90	385	1.19	1.29	—	0.64	0.60	—	278±8	276	5.16	—	1.8	2.2	+4.2	+5.8	+1.2
10	22.2	55.6	22.2	4.40₆	2.50	1.00	345	1.28	1.28	—	1.14	0.32	—	200±8	215	5.12	—	1.8	2.1	+1.4	+2.7	-1.9
11	20.6	60.0	20.0	4.405	3.00	1.00	384	1.14	1.14	0.80	0.74	—	—	188±8	194	4.56	(2.40)?	1.8	2.1	+1.4	+2.7	-1.9
12	26.6	60.0	13.4	4.45	2.26	0.50	290	1.54	0.77	—	0.38	—	—	175±5	177	3.08	—	1.8	2.0	+2.1	+3.5	-1.1
13	13.4	60.0	26.6	4.37	4.48	1.99	573	0.76	1.51	—	—	0.19	—	229±4	217	6.04	—	1.8	2.2	+2.4	+3.6	-1.0
14	29.6	55.6	14.8	4.439	1.88	0.50	260	1.71	0.86	—	—	0.10	—	184±3	184	3.44	—	1.8	2.0	+2.2	+3.7	-0.9
15	14.8	55.6	29.6	4.39₁	3.73	1.99	514	0.85	1.59	—	1.51	0.43	0.10	235±3	244	6.36	—	1.8	1.8?	+0.4	+1.6?	-3?

uniformly atomically dispersed mixtures of AsSe$_{3/2}$ and SeGe$_{4/2}$ s.u. and may entail the formation of chemically micrononuniform structure in the glass as a result of colloidally dispersed differentiation of the latter in the path of the structural units. In this case, some of the valence electrons remain unpaired at the (AsSe$_{3/2}$)$_m$ and (GeSe$_{4/2}$)$_n$ microphase boundary as a result of the spatial mismatch of the valence bond periodicity. Cul-de-sacs (or traps) for the current carriers are created, thus limiting the through conductivity. In this case we should expect to find lower values for the steric factor: log β < 0 [4]. The experimental conductivity data clearly confirms these views.

2. The chemical compositions, from the initial batch compositions, of the AsSe$_x$Ge$_y$ glasses in question are shown in Table 1. Germanium (n-type; 54 Ω·cm) and double distilled arsenic and selenium were used as raw materials.

From the chemical composition data in Table 1 the approximate concentrations of the GeSe$_{4/2}$: AsSe$_{3/2}$·AsAs$_{3/3}$ and GeGe$_{4/4}$ structural units are derived. It is assumed that selenium reacts primarily with germanium and secondarily with arsenic. The decrease in free energy in the analogous transfer of oxygen from arsenic to germanium [6] is an indication in favor of this assumption:

$$As_4O_6 + 3Ge \rightarrow 2GeO_2 + 4As, \quad F_{25°C} = -106$$

The preferred combination of selenium with germanium is in accordance with the electronegativity series (X$_{Ge}$ = 1.8; X$_{As}$ = 2.0; and X$_{Se}$ = 2.4) from which it follows that the ionic term in the chemical bond Δ$_{A-B}$ = 23(X$_A$ − X$_B$)^2 is larger in Ge−Se (Δ ≈ 8 kcal/mole) than in As−Se (Δ ≈ 4 kcal/mole) [7].

Unfortunately we do not have available reliable data for the energy of the individual covalent bonds ε$_{Ge-Ge}$, ε$_{As-As}$, ε$_{Se-Se}$. Their average value is 41 ± 4 kcal/mole-bond, derived from atomization energies (39, 40, and 48 kcal/mole-bonds respectively according to Latimer's thermochemical data in [6]; 45, 35, and 41 kcal/mole according to Cottrel's [8]; and 38, 32, and 44 kcal/mole from Pauling's [7]). Taking the arithmetical average value of the covalent bond energy and making allowance for the values above of Pauling [7] for the ionic components of these bonds, we obtain the values for the bond energy $\varepsilon_{Ge-Se} \approx 49$ kcal/mole and $\varepsilon_{As-Se} \approx 45$ kcal/mole–bond. It follows that

$$4AsSe_{3/2} + 3Ge \rightarrow 3GeSe_{4/2} + 4As - 48 \text{ kcal.}$$

These results, as well as the calculation below for the concentrations of structural units, are tentative. The rough averaging, necessary in view of the diversity of the initial data for the energy of the individual covalent bonds in germanium, arsenic and selenium, and the absence of data which permits us to allow for the temperature dependence of the undoubtedly essential equilibrium in the glass melting process, indicate the urgent and basic necessity for appropriate experimental thermodynamic studies in the field of semiconducting materials.

3. The calculation of the concentration of structural units is shown schematically in Table 2. It was assumed that the condition: $x-2y \le 1.5$ (absence of cycles and chains in selenium) was satisfied in these $AsSe_x Ge_y$ glasses. We used as a basic parameter the arsenic concentration [As] mole/cm^3, equal to [M], where M is $AsSe_x Ge_y$. Then [M] = [$AsSe_x Ge_y$] = d/M = [As]; where d is the density and M is the conventional molecular weight of $AsSe_x Ge_y$.

From Table 2 it follows that the glasses in question of variable composition form four groups which are distinguished by the structural-unit concentrations.

I. This group of glasses contains enough selenium for the unpaired electrons of its atoms to combine fully with the unpaired electrons of the germanium and arsenic atoms and form electron-pair covalent bonds which complete the octet of the valence electron shells. From the laws of stoichiometry we find that for the Group I $AsSe_x Ge_y$ glasses $x = 1.5 + 2y$. As a result, the concentrations of [$GeSe_{4/2}$] = y[M] and [$:AsSe_{3/2}$] = [M] in such glasses.

II. This group satisfies the condition $1.0 \le x-2y \le 1.5$. Following the argument above it is assumed that selenium reacts completely with the germanium and with the arsenic produces not only $:AsSe_{3/2}$, but also structural units depleted in selenium. As there is insufficient selenium to convert all the arsenic to $As_2Se_{4/2}$, structural units such as $\begin{matrix} -Se \\ -Se \end{matrix} \! \! As-As \! \! \begin{matrix} Se- \\ Se- \end{matrix}$, or $AsSe_{2/2}$ appear. It is clear that as before [$GeSe_{4/2}$] = y[M]; further [$:AsSe_{3/2}$] = 2(x−2y−1) [M] and [$\cdot AsSe_{2/2}$] = 2[1.5−(x−2y)] [M].

III. This group of glasses is so depleted in selenium that the arsenic remains unreacted with the selenium.

TABLE 2. Calculation of the Concentration of Chemical Structural Units in $AsSe_x Ge_y$ Glasses, in Mole/cm^3

Structural units	Types of atomic proportion relations			
	I $x-2y=1.5$	II $1.0 < x-2y < 1.5$	III $0 < x-2y < 1.0$	IV $x-2y < 0$
$GeSe_{4/2}$	y [M]	y [M]	y [M]	0.5x [M]
$:AsSe_{3/2}$	[M]	2 (x−2y−1)[M]	—	—
$\cdot AsSe_{2/2}$	—	2[1.5−(x−2y)]·[M]	(x−2y) [M]	—
$AsAs_{3/2}$	—	—	0.5[1−(x−2y)]·[M]	0.5 [M]
$GeGe_{4/4}$	—	—	—	0.5(y−0.5x)[M]

IV.   In this group of glasses the selenium is not quite sufficient to satisfy the Ge and therefore $[GeSe_{4/2}] = 0.5x\,[M]$. The remaining expressions in Table 2 for the concentration of various structural units are readily solved algebraically.

4.   The experimental data for the microhardness H in Table 1 was obtained using a PMT-3 instrument. The measurements were made on glass specimens after the conductivity measurements. The specimens were polished with GOI paste to a mirror finish. The optimum loading was 50 g (except for glasses Nos. 7 and 9 where it was 100 g). With such loading the diagonal dimensions of the prismatic impressions were from 20-40 μ and as a rule no cracks were observed.

Microhardness results depend on many factors which are not always controllable and are therefore quite often poorly reproduced. Not only by different laboratories but in independent determinations, a significant spread in the results of different workers in the same laboratory may occur. It is therefore not unexpected when we find marked variations in the values of the molar-volume microhardness h_i in the expression for microhardness

$$H = \sum h_l\,[x_i],\qquad(2)$$

where $[x_i]$ is the molar-volume content of the x_i structural units. Thus from the experimental data for microhardness H, different workers have obtained the following values for $h_{:AsSe_{3/2}}$: $6.6 \cdot 10^3$ [9], $5.9 \cdot 10^3$ [1], and $6.4 \cdot 10^3$ kg/mm$^2 \cdot$cm^3 mole s.u. [10] and for $h_{GeSe_{4/2}}$: $12.1 \cdot 10^3$ [9], and $8.8 \cdot 10^3$ kg/mm$^2 \cdot$cm^3/mole s.u.

In this work we chose the partial molar volume microhardness $h_{:AsSe_{3/2}}$ as the basic parameter which is taken as $6.0 \cdot 10^3$ kg/mm$^2 \cdot$cm^3/mole s.u. as there is satisfactorily agreement by various authors on this value. From this value, with the addition of the new experimental data for the microhardness H in the group I glasses No. 1-5 (x-2y = 1.5; Tables 1 and 2) and using Eq. (2), we determined the average value $h_{GeSe_{4/2}} \approx 12 \cdot 10^3$ kg/mm$^2 \cdot$cm^3/mole s.u. (which agrees satisfactorily with the data of L. G. Aio and V. F. Kokorina [9]). By using the values of $h_{AsSe_{3/2}}$ and $h_{GeSe_{4/2}}$ and the experimental data for microhardness in the Group II, (Nos. 11 and 12 glasses (Table 1)), as a basis we determined analogously the value of $h_{(.AsSe_{2/2})} \approx 5 \cdot 10^3$ kg/ mm$^2 \cdot$cm^3/mole s.u.

In the same way we subsequently dealt with the Group III glasses (Nos. 6, 8, 10, 13, and 14), Group IV (Nos. 7, 9) and determined the values $h_{AsSe_{3/3}} \approx 9 \cdot 10^3$ and $h_{GeGe_{4/4}} \approx 14 \cdot 10^3$ kg/mm^2 cm^3/mole s.u.

The additive relationship of Eq. (2) on which these calculations were based is found in practice to be satisfactory. This is borne out by the comparison in Table 1 between the experimental values for the microhardness H and those calculated from Eq. (2) with the addition of the values for h and the concentrations of the s.u. (Table 1). The mean difference of 4% between the calculated and the experimental values of H is reasonable considering the 3% mean experimental error.

The microhardness results support the views outlined above concerning the structural chemical nature of the glasses in question.

5.   The conductivity of the glasses was measured by an electrometric method [1] using graphite electrodes. Silver paste contacts did not give satisfactorily reproducible conductivity-temperature curves for the $AsSe_xGe_y$ glasses.

Figure 1 shows the data for log $\sigma = f(10^4/T)$. The values of the energy ε_0 and log σ_{0e} derived from the graph are listed in Table 1. The quantum energy ε_λ for optical ionization of the covalent bond, corresponding to the light absorption edge determined using a SF-5 spectrometer, is also given in Table 1. The electron-pair concentration [v] used in the calculation of

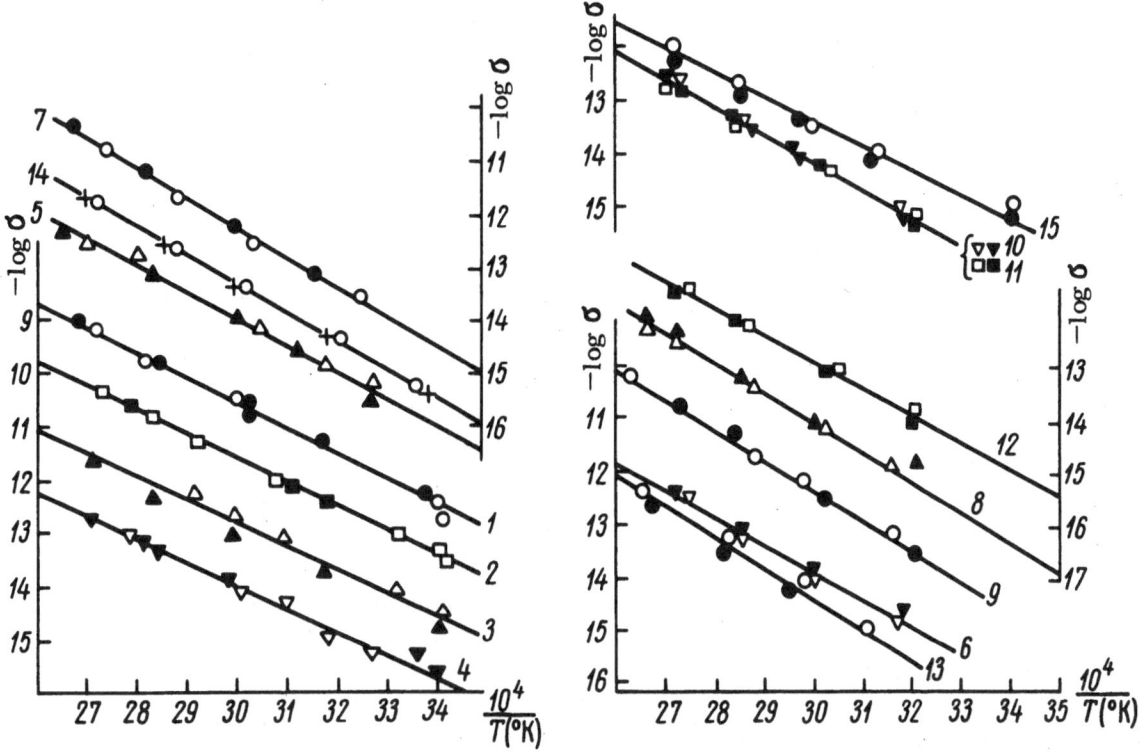

Fig. 1. Logarithmic temperature dependence of the specific electrical conductivity. The numbers of the plots refer to the sample designations in Table 1. The geometric symbols distinguish the melts from which the samples came.

the conductivity modulus log σ_{0e} / [v] was equated to the concentration of the majority type of bonds in the glass which provide the through transport of current carriers. As can be seen from Table 1, these bonds were Ge—Se and As—Se belonging to the $GeSe_{4/2}$ and $AsSe_{3/2}$ structural units. The remaining structural units were, as a rule, blocked. The higher $AsSe_{2/2}$ concentration in glass No. 14 did not produce any significant change in the energy value or the conductivity modulus.

Glass No. 1, corresponding to arsenic selenide containing 10% germanium selenide, had a bond ionization energy $\varepsilon_i \approx \varepsilon_\lambda = 1.7$ eV and activation energy $\varepsilon_a = \varepsilon_\sigma - \varepsilon_i/2 \approx 0.05$ eV which are characteristic of $AsSe_{1.5}$ [1]; log $\beta = -0.2$ corresponded to through conductivity in accordance with the valence theory in [4].

In glasses Nos. 2 and 3 with a majority of As—Se bonds, the values of ε_λ and ε_σ are preserved. The rise in the $GeSe_{4/2}$ concentration consequent on the fall in $AsSe_{3/2}$ was accompanied by a decrease in the value of the log β term caused by the growing destruction of the periodicity in the network and by the associated limitation of the through conductivity. This phenomenon was also noted in glass No. 4. In the latter, in spite of the closeness of the [Ge—Se] and [As—Se] values, the conductivity is determined by the As—Se bond ionization energy which is lower. In fact, when the concentration of the two types of bonds is equal a consequent excess conductivity will be observed:

$$\frac{\sigma_{As-Se}}{\sigma_{Ge-Se}} \approx \exp\left(-\frac{\varepsilon_{\sigma_{As-Se}} - \varepsilon_{\sigma_{Ge-Se}}}{2kT}\right) \approx \exp\left[-\frac{(1.7-1.8)-(2.1-2.2)}{2kT}\right] \approx 10^3 - 10^4$$

at 300°K. This value is noticeably decreased by the steric term β (discontinuities in bond skeleton in the $AsSe_{3/2}$ and $GeSe_{4/2}$ mixtures). The conductivity of glass No. 5 is to a considerable extent determined by the Ge−Se bonds and this is borne out by the values of ε_λ and ε_σ [2]. The slight increase in the β term from glass No. 4 to No. 5 corresponds to a decrease in the number of incidences of breakdown in the periodicity of the bond network due to the appearance of a majority of Ge−Se bonds.

Glasses of Group II (Nos. 11 and 12; $\varepsilon_\lambda = 1.8$; $\varepsilon_\sigma = 2.0-2.1$ eV), Group III (Nos. 6, 8, 10, 13, and 14; $\varepsilon_\lambda = 1.8$; $\varepsilon_\sigma = 2.0-2.2$ eV), and Group IV (Nos. 7 and 9; $\varepsilon_\sigma = 2.2$ eV) are characterized by the presence of a majority of Ge−Se bonds and the conductivity is determined by the ionization of these bonds. Glass No. 15 occupies a peculiar position and requires further study.

In glasses Nos. 6, 10, and 11 with increased $\cdot AsSe_{2/2}$ concentration, the value of the steric term $\log \beta \approx -2$ is attributable to the break-down of the periodicity of the bond network. Glasses Nos. 12 and 14 with $\log \beta \approx -1$ represent the transition to glasses Nos. 7-9 and 13 with a steric term $\log \beta = \pm 1$ which indicates that there is a continuous valence bond network providing for the through transport of activated current carriers.

An essential feature of the glasses Nos. 7-9 is the presence of a significant concentration of excess arsenic and germanium, although they do not decrease the energy $\varepsilon_\sigma = 2.0$ to 2.2 eV which corresponds to an excess concentration of Ge−Se bonds. At the same time in Nos. 7 and 9 the excess of germanium is clearly present in a crudely dispersed state which prevents the spectrophotometric measurement of an optical absorption edge. It follows that the relatively high conductivity of the structured particles of germanium ($\varepsilon_\sigma \approx 0.7$ eV) and arsenic ($\varepsilon_\sigma \approx 1.2$ eV) [11] was suppressed by the blocking vitreous $GeSe_{4/2}$ network which limited current carrier transport. This phenomenon in glasses does not conform to the well defined effect of small amounts of impurities on the conductivity of simple crystals of a range of semiconductors. The structural chemical nature of foreign inclusions cannot be considered to be completely clear.

Thus, the conductivity data for vitreous $AsSe_xGe_y$ demonstrates that conduction is limited by the mobility of the current carriers in the skeletal network of the predominant covalent bonds in the glass. Inclusions of foreign structures are blocked and thus have no significant effect on the conductivity.

The validity of the "valence hypothesis" of the conductivity of glass has been clearly demonstrated and the use of structural chemical concepts in vitreous solids is clearly fruitful.

Summary

1. The composition and concentrations of the structural chemical units in $AsSe_xGe_y$ glasses of various compositions were provisionally determined.

2. The microhardness of $AsSe_xGe_y$ glasses has been measured, the molar values for the microhardness of the structural units determined, and the additive functional dependence of the microhardness of glasses on their structural unit concentration has been established.

3. The conductivity of $AsSe_xGe_y$ glasses and the values of the energy ε_σ and the modulus of conductivity have been determined. We have established that the conductivity of these glasses is determined by the skeletal network of the majority type of covalent bonds, in keeping with the valence nature of conductivity.

References

1.   L. A. Baidakov, Z. U. Borisova, and R. L. Myuller, Zh. Prikl. Khim., 34:2446 (1961)
      ● this volume, p. 133.
2.   Z. U. Borisova, R. L. Myuller, Chin. Ch'eng-ts'ai, Zh. Prikl. Khim., 35:774 (1962).
3.   R. L. Myuller and T. P. Markova, Vestn. Leningr. Gos. Univ., Vol. 17: No. 4, Iss. No. 1, p. 75 (1962).

4. R. L. Myuller, Vestn. Leningr. Gos. Univ., Vol. 16, No. 22, Iss. 4, 86 (1961); in: Physics, Proceedings of the XXth. Scientific Conference of the L. I. S. I., Leningrad (1962), p. 18.

5. R. L. Myuller, in: The Structure of Glass, Vol. 2, Consultants Bureau, New York (1960), p. 50, ● this volume, p. 178.

6. W. M. Latimer, The Oxidation State of the Elements and their Potentials in Aqueous Solution, 2nd ed., Prentice-Hall, New York (1952).

7. L. Pauling, The Nature of the Chemical Bond, 3rd ed., Cornell University Press, New York (1960).

8. T. L. Cottrell, The Strength of Chemical Bonds, London (1954).

9. L. G. Aino and V. F. Kokorina, Opt-mekh. Prom., 28(6):48 (1961).

10. A. V. Danilov and R. L. Myuller, in: Physics, Proceedings of the XXth Scientific Conference of the L. I. S. I., Leningrad (1962), p. 21.

11. T. S. Moss, Photoconductivity of the Elements, London (1962).

A Study of the Crystallization of $AsSe_xGe_y$ Glasses from Conductivity Measurements*

1. The practical need for partially crystallized vitreous materials as well as for glasses which are free from crystalline inclusions emphasizes the importance of experimental and theoretical studies of the crystallization kinetics of glasses, a subdivision of the study of crystallization from the melt. Similar investigations play an essential role in establishing the origin of the defects in crystalline solids which affect the electrochemical, optical and other physicochemical properties of various materials and are very important to contemporary science and technology. Such studies are also valuable in solving the problem of glass formation and in understanding the causes of low crystallization tendency in groups of materials of particular chemical compositions.

Tamman's classical work and later studies on the crystallization of melts [1-3] have qualitatively demonstrated the nature of the basic physical processes in the development of crystallization centers and in subsequent crystal growth. The number of crystallization centers in the volume of the glass at a given time is determined by the temperature dependence of the rate of formation of nucleation centers and the time spent in a particular temperature range during the cooling of the solidifying melt. The growth rate of the crystals has an analogous type of temperature-time dependence.

The quantitative aspect of the kinetics of glass crystallization is a relatively poorly developed scientific field. There has been some qualitative study of the morphology of crystals developing in complex technical glasses, of the linear growth rate of crystals and sphaerolites, and of crystal growth models using, to some extent, conventional mathematical methods [2, 4-8]. Little attention has been paid to the structural chemical nature of the melt and to the role in crystallization of primary chemical processes. Several papers [9-11] in recent years have considered this question. We shall consider, principally, structural chemical reorganizations involving the transformation of chemical bonds (largely covalent) which require a significant activation energy [12]. The latter is noticeably less than the corresponding energy for bond breaking which in its turn cannot serve as a criterion of glass formation. The difference in principle between the points of view expressed by Sun [13] and by us in [12] is expressed in the contrasting idea we have developed of localized short-range covalent bonds between atoms (requiring a higher activation energy for displacement) and central long-range coulombic bonds (requiring less activation energy for displacement).

2. In any vitreous material in the critical temperature region (below the liquidus temperature and above the region in which the valence degrees of freedom of the atomic vibrations are frozen) there is a particular temperature-range in which nucleation centers form and crystals grow at significant rates. The significance of supercooling this melt is clear.

*R. L. Myuller and E. V. Shkol'nikov, Vestn. Leningr. Gos. Univ., No. 22, p. 119 (1962).

Below the optimum temperature range for crystallization the nucleation and crystal growth rates decrease rapidly as a result of the exponential decrease in the probability of activation of the structural units of the vitreous network. Activation is required for realizing the transformation of the covalent bonds during the regrouping of the atoms necessary for the transition from the disordered structure of the vitreous network to the ordered structure of the crystalline lattice [12]. In this case, in addition to the local bond restructuring necessary to create the particular atomic configurations with a minimum chemical potential in glasses of a complex composition (in particular in those deviating from stoichiometric ratios), diffusive interchange of some of the structural units is required [4, 14, 15]. This combination of kinetic processes is ultimately limited, in the same way as the viscous flow of glass [16], by the transformation rate of the covalent bonds. It is therefore reasonable that the maximum crystallization rate of a glass is linearly dependent on the reciprocal of the viscosity [1, 4, 6, 12, and 15].

The activated processes, which obey chemical kinetic laws, can be most readily studied kinetically at temperatures below the temperature of the maximum crystallization rate. The study, therefore, of the rate of slow uniform crystallization throughout the volume of the glass deserves some serious considerations.

3. The ionic conductivity of sodium silicate glasses falls significantly during volume crystallization [17]. According to the results of Mazurin and Tsekhomskii [18], completely crystallized silicate glasses have significantly higher energies of conduction than the glass from which they were derived. Mazurin's result shows that the preexponential term σ_0 retains its value, and as a calculation of the conductivity modulus $L = \log (\sigma_0/[Na])$[19] shows (for sodium metasilicate $L = 2.9 \pm 0.1$; for disilicate $L = 4.0 \pm 0.4$; theoretical value $L = 3.7 \pm 1$), this indicates that all the ions are involved in the statistical distribution on electrolytic dissociation. No one apparently has attempted to calculate the kinetics of the crystallization rate constant on the basis of the conductivity changes.

An increase in the ionic conductivity of glass on crystallization is due to an increase in the dissociation energy of the polar units. A more condensed atomic packing occurs during crystallization and produces a closer approach of the electron shells of the atoms and ions. As a result, when the sodium ion escapes from the polar unit into the interstitial space an increased energy is required after crystallization to overcome the increased electron shell repulsive forces in the interstices. The decreased autosolvation energy of dissociated sodium ions and the increase in the free energy of dissociates of the polar units is due to this same effect.

The electrons, being incommensurably smaller than the ions, do not suffer similar spacial difficulties on the ionization of covalent bonds in crystallized glasses. The rise in the conductivity and the fall in the conductivity energy ε_σ [21, 22] on the crystallization of electron-conducting glasses indicates that there is then a significant decrease in the ionization energy of the covalent bonds.

In their work on the conductivity of GeSe$_{1.5}$ glasses containing 40 at.% of germanium with 20 at % of the selenium replaced by arsenic and crystallized to various degrees, Borisova, Shkol'nikov, and Kozhina established that the energy of conduction of such glasses decreased on crystallization, and that the conductivity increased by 7–10 orders of magnitude. The emergence of a crystalline phase of GeSe was thus established in [23].

A more detailed study of crystallization in these glasses is of considerable interest since it may provide an opportunity to establish more exactly the temperature–time characteristics of this process.

The feasibility of such a time-consuming study was also suggested by the satisfactory reproducibility of the conductivity values for partially crystallized glasses reported by Borisova et al. in [23]. The most significant fact was that, in this case, the kinetics of crystallization was followed for glasses with an atomic structural network connected by clearly covalent chemical bonds. We should expect to obtain valuable information on crystallization rates which were determined by the activation of localized covalent bonds in the absence of ionic components.

4. In order to study the kinetics of volume crystallization by conductivity measurements we have adopted the method first used by Lebedev [15, 24]. A specimen was heated at an appropriate temperature for a known time and then quenched rapidly. The conductivity of the quenched specimen was measured at such a low temperature that the state of crystallization was completely frozen. A small temperature dependence of conductivity was taken as characterizing the degree of crystallization of the heat-treated glass specimens.

It is known that the predominance of surface crystallization in glass is initiated by the presence of surface active impurities and the products of atmospheric oxidation, as well as by the directive forces of surface tension and by the loss of individual volatile components. In order then to stimulate volume crystallization dispersed additives can be used, the so-called crystallization catalysts. Cases are also known of successful volume crystallization without the use of additives, for example the crystallization of lithium silicate glasses, of glasses in the alumino-borosilicate system, and vitreous selenium [11, 25]. In this work we have studied the volume crystallization of vitreous $AsSe_{5.0}Ge_{4.0}$ and $AsSe_{3.0}Ge_{2.66}$ (corresponding to the glasses $GeSe_{1.25}As_{0.25}$ and $GeSe_{1.125}As_{0.375}$ obtained in [23]) without added catalyst.

The starting materials for the syntheses were: Grade V-4 and V-3 Se; GP (n-type 54 $\Omega \cdot$cm) Ge; and double distilled As. The evacuated quartz ampoules containing the batch were heated to 600° at a rate of 250-300°C per hour and kept at that temperature for 1 h. The temperature was then raised to 900° at an average rate of 150° per h. The ampoules were kept at 900°C for 5-8 h in order to homogenize the melt. They were then quenched by cooling to room temperature in 10 min.

The vitreous state of the initial specimens was confirmed visually by the conchoidal fractures, by the absence of lines on x-ray powder photographs, by the absence of inclusions when examined with an MIK-1 metallurgical microscope, and by light transmission with an IKS-12.

The density was determined by the Archimedean method. The microhardness was measured with a PMT-3 instrument with a diamond pyramid with an angle of 136° and loads of 50 and 100g. The low temperature conductivity was determined by an electrometric method using a platinum guard ring and graphite contacts. This method allowed us to measure the resistance of specimens from 10^2 to $10^{17}\Omega$ [26, 27]. The temperature variations during the conductivity measurements did not exceed ± 0.5°C.

The quenched specimens of the initial glasses obtained in this way (in the form of disks weighing ~5 g) were washed in ethyl alcohol, wrapped in aluminum foil and placed in a quartz or molybdenum glass ampoule. The ampoule was wrapped in moist asbestos (to keep it cool during sealing) evacuated to 10^{-3} mm Hg, and sealed. The heat treatment of the glass specimens during volume crystallization was carried out at a constant temperature, at the low temperature end of the optimum crystallization range in the critical temperature region [16, 28]. The temperature range was established empirically for each glass composition. In this case the rate of emergence of volume crystallization centers was clearly commensurate with the rate of formation at the glass surface.

Two ampoules containing specimens were placed in a vertical tubular furnace at the required temperature. The furnace temperature was automatically maintained within ± 1° up to 300°C by a contact thermometer, above 300°C within ± 4-5° by an ÉPV-11A automatic potentiometer and then by the method described by Alekseev [27]. The temperature was measured with an M-198/3 using a calibrated KhK thermocouple, the hot junction of which was in contact with

TABLE 1. Characteristics of the Initial Glasses and Crystalline GeSe

Glass No.	I	II	—
Chemical composition of the unit M	AsSe$_{5.0}$Ge$_{4.0}$	AsSe$_{3.00}$Ge$_{2.66}$	GeSe
Molecular weight of M, arbitrary units	760	505	152
Density d, g/cm^3, of vitreous → crystalline (in brackets is the accepted limiting value, d$_\infty$) specimens	4.49 → 4.92 (4.95)	4.59 → 5.07 (5.10)	5.3
Glass I at 399 ± 4°C — M	5.9 → 6.5	9.1 → 10.0	35.0
As—Se	2 [M] = 11.8 → 13.0	0.68 [M] = 6.2 → 6.8	—
Ge—Se	4 [M] = 23.6 → 26.0	2.66 [M] = 24.2 → 26.6	[M] = 35.0
All atoms	10 [M] = 59.0 → 64.8	6.66 [M] = 60.5 → 66.9	2 [M] = 70.0
Microhardness, H, kg/mm^2	242 ± 7 → 87 ± 8	308 ± 8 → 76 ± 4	86 ± 9
Conductivity energy σ_0, eV vitreous → crystalline (the accepted limiting value ε_∞ is in brackets)	1.50 → 0.33 (0.30)	1.70 → 0.47 (0.40)	0.28
log δ_{0e} (initial)	—0.9	1.7	0.1

the ampoule wall in the vicinity of the specimen while the cold junction was in melting ice. Sublimation of selenium was insignificant during the heat treatment in the evacuated ampoule (maximum pressure 5 mm Hg [25]).

The glass specimens, after the appropriate heat treatment, were quenched in the ampoule to room temperature in 2-4 min. After removal from the ampoule, the specimens were ground to remove the oxide film the thickness of which was less than 0.05 mm. We then measured the density, microhardness, and low temperature conductivity. X-ray structural analyses were made by the powder method, using a URS-70 with a Cu anode.

5. Table 1 shows the values of the density, microhardness, and the other parameters which characterize the quenched glasses I and II in the initial state and in the final crystalline state. The limiting values of the density d$_\infty$ and energy ε_∞ used in the calculations differ little from the values at the completion of crystallization. The calculated concentrations of the [As—Se] and [Ge—Ge] covalent bonds are also given here. It follows from the data in Table 1 that there is an increase of 10-12% in the packing density of the atoms [A] during the crystallization of the glasses.

Figures 1 to 3 show the experimental data for the low temperature dependence of the conductivity of Glasses I and II in the initial state and after lengthy heat treatment in stages according to Table 2. Figures 1-3 show the satisfactory reproducibility of the data from glasses of parallel melts, indicating the insignificant effect of impurities.

Table 2 gives the values of the energy ε_σ and log σ_{0e} calculated from the data in Figs. 1-3. According to the latter the familiar law $\sigma = \sigma_{0e} \exp(-\varepsilon_\sigma/2kT)$ is satisfied in all cases.

In the first stages of heat treatment we observed an induction period in which the density and microhardness were practically unchanged while the energy ε_σ and log σ_{0e} increased. Figure 4 illustrates the subsequent changes in these parameters in Glass I at 399°C and II at 450°C. After

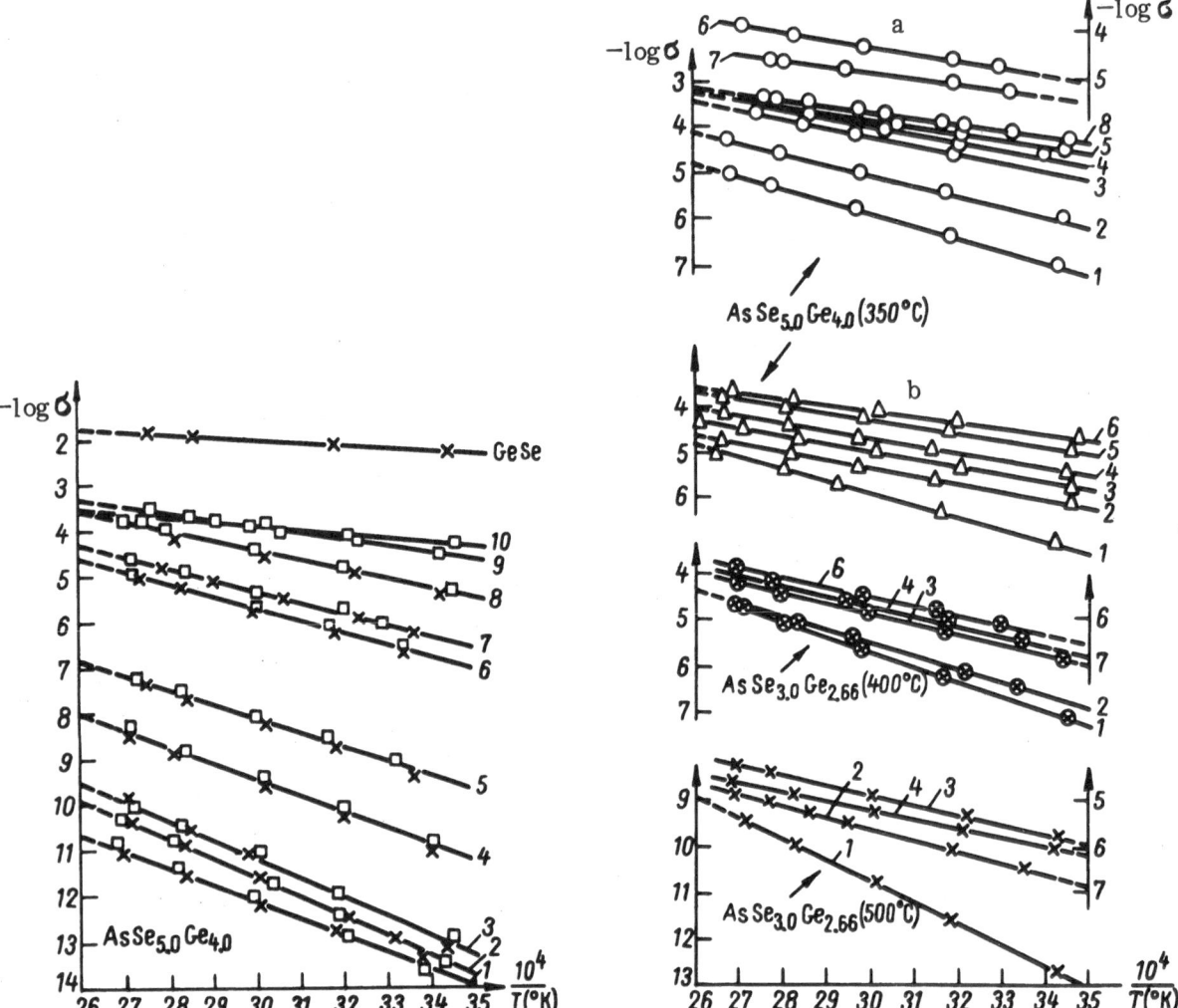

Fig. 1. The temperature dependence−log σ = $f(10^4/T°K)$ in Glass I (AsSe$_{5.0}$Ge$_{4.0}$) directly after quenching (stage 1) and after heat treatment at 300°C in stages 2−10 according to Table 2. The different symbols correspond to glass specimens from two parallel melts. The straight line GeSe corresponds to GeSe polycrystalline aggregate.

Fig. 2. The temperature dependence−log σ = $f(10^4/T°K)$ in Glass II (AsSe$_{3.0}$Ge$_{2.66}$) after treatment at 400 and 500°C. As in Glass I (AsSE$_{5.0}$Ge$_{4.0}$) after heating at 350°C in stages with a prepatory heat treatment at a) 399°C and b) 450°C according to Table 2.

the induction period has been completed, the parameters H, d, and ε_σ change monotonically and asymptotically approach the limiting values H_∞, d_∞, and ε_∞. An analogous behavior was observed by Tykachinskii and Sorkin [29] and has previously been discussed by Frenkel' [30], Pines [31], and Gorskii [32].

In order to accelerate the process of crystallization during the induction period which occurred at the lower temperature (350°C), we used a specimen of Glass I preheated at 399°C and another at 450°C (see Table 2 and Fig. 2). The subsequent analysis (the significant spread in the values of P_d and P_e) indicated the distorting effect of preheating at a higher temperature on the results of the subsequent heat treatment (as in [33] the corresponding results were omitted in calculating the activation energy of crystallization of the glass).

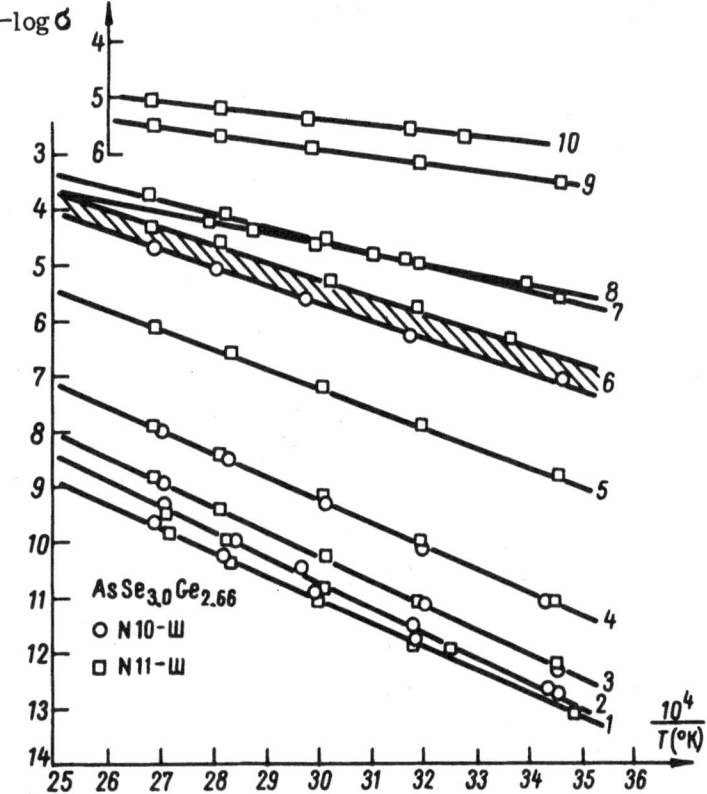

Fig. 3. Temperature dependence - log $\sigma = f(10^4/T°K)$ in Glass II (AsSe$_{3.0}$Ge$_{2.66}$) after quenching (stage 1) and after heat treatment at 450°C in stages 2 to 10 according to Table 2. The different symbols correspond to specimens from two parallel melts.

At the beginning of the heat treatment the x-ray powder photographs of the glasses showed faint lines which increased in intensity after heating. The lines corresponded to crystalline GeSe with a distorted rhombic NaCl-type lattice [23, 24]. Crystalline arsenic was not detected. In crystalline GeSe one finds, in addition to the Ge−Se ($\varepsilon_\sigma \approx 2.0$-$2.2$ eV) covalent bonds which are ionized with difficulty [35], the easily ionized Ge−Ge bonds [36, 37]. It is therefore reasonable that there should be a significant drop in the conductivity energy during the crystallization of AsSe$_{5.0}$Ge$_{4.0}$ and AsSe$_{3.0}$Ge$_{2.66}$ glasses when there is formation of crystalline GeSe. For example, in the heat treatment of Glass I (399°C) ε_σ falls from 1.7 ± 0.1 eV corresponding to the As−Se bond ionization [38, 39] to 0.33 ± 0.02 (see Table 2) corresponding to the value of 0.28 ± 0.03 eV for GeSe. At the same time we noted a decrease in the microhardness to the values observed for GeSe.

We should particularly take note of the fact that the enormous rise in conductivity (by 7 to 10 orders of magnitude) during the crystallization of these glasses is due to the decrease, which we have just discussed, in the bond ionization energy. The value of log σ_{0e} does not in this case undergo such fundamental changes.

The experimental values of the conductivity modulus log (σ_{0e}/[As−Se]) and log (σ_{0e}/[Ge−Ge]) are close to the theoretical values and log β parameter [19] is generally within the theoretical limits of log $\beta = \pm 1$ (see Table 2). A systematic decrease in the values of log $\beta < -1$ is observed only in the initial quenched glasses and at the end of crystallization. This is most probably explained by the presence in the melt of cyclic structures and of ruptured bonds which

TABLE 2. Experimental Data for Isothermal Volume Crystallization

Stage of heat treatment (Figs. 1-3)	Time, h	Density d, g/cm³	Microhardness H, kg/mm²	ε_g, eV	$\log \sigma_{0e}$	$\log \beta$ As—Se	$\log \beta$ Ge—Ge
Glass I at 399 ± 4°C							
1	0	4.492	242±7	1.50	−0.9	−3.6	−4.1
2	0.5	4.489	–	1.70	1.1	−1.6	−1.9
3	1.0	4.50	253±10	1.70	1.5	−1.2	−1.5
4	1.25	–	–	1.50	1.9	−0.8	−1.1
5	1.50	4.55	–	1.35	2.1	−0.6	−0.9
6	2.00	4.66	238±10	1.09	2.5	−0.2	−0.5
7	2.50	4.69	–	1.03	2.4	−0.3	−0.6
8	10	4.77	122±11	0.81	1.6	−1.1	−1.4
9	18	4.86	87±8	0.55	0.2	−2.5	−2.8
10	31	4.92	–	0.33	−1.5	−4.2	−4.8
Crystalline GeSe		5.3	86±9	0.28	0.1	–	−3.1

Glass I at 350 ± 4°C (preheated at 399 ± 4°C)

Stage	Preheat	Time, h	Density d, g/cm³	Microhardness H, kg/mm²	ε_g, eV	$\log \sigma_{0e}$	$\log \beta$ As—Se	$\log \beta$ Ge—Ge
–	399 ±4°C	1	4.50	253	1.70	1.5	−1.2	−1.5
1		2	4.67	238	1.08	2.2	−0.5	−0.8
2		4	4.72	–	0.94	2.0	−0.7	−1.0
3	350 ±4°C	12	4.77	112±10	0.81	1.8	−0.9	−1.2
4		17	4.79	102±12	0.71	1.4	−1.3	−1.6
5		23	4.82	–	0.65	1.1	−1.6	−1.9
6		35	4.85	–	0.62	0.5	−1.2	−2.5
7		42	4.86	–	0.57	−0.5	−3.2	−3.5
8		47	4.86	88±9	0.55	−0.4	−3.1	−3.4

Glass I at 350 ± 4°C (preheated at 350 ± 4°C)

Stage	Preheat	Time, h	Density d, g/cm³	Microhardness H, kg/mm²	ε_g, eV	$\log \sigma_{0e}$	$\log \beta$ As—Se	$\log \beta$ Ge—Ge
1	350 ±4°C	1	4.67	253	1.07	2.1	−0 6	−0.9
2		3	4.72	–	0.75	0.3	−2.4	−2.7
3		5	4.76	–	0.71	0.4	−2.3	−2.6
4	450 ±5°C	9	4.78_2	–	0.64	0.1	−2.6	−2.9
5		12	4.78_5	–	0.59	0.25	−2.4	−2.7
6		14	4.79_0	90±8	0.57	0.15	−2.5	−2.8

Glass II at 400° ± 4°C

Stage	Time, h	Density d, g/cm³	Microhardness H, kg/mm²	ε_g, eV	$\log \sigma_{0e}$	$\log \beta$ As—Se	$\log \beta$ Ge—Ge
–	0	4.591	308±8	1.70	1.7	−0.9	−1.3
–	2	4.60	278±3	1.85	3.1	0.5	0.1
1	8	4.78	128±8	1.24	3.8	1.2	0.8
2	12	4.85	–	1.08	2.6	0	−0.4
3	20	4.92	–	0.93	1.2	−1.4	−0.8
4	32	4.95	–	0.86	0.7	−1.9	−2.3
5	42	4.97	–	0.82	0.5	−2.1	−2.5
6	52	4.98	76±4	0.78	0.4	−2.2	−2.6

Glass II at 500 ± 6°C

Stage	Time, h	Density d, g/cm³	Microhardness H, kg/mm²	ε_g, eV	$\log \sigma_{0e}$	$\log \beta$ As—Se	$\log \beta$ Ge—Ge
1	0	4.59	308	1.70	1.7	−0.9	−1.3
–	2	4.60	278±3	1.85	3.1	0.5	0.1
2	3	4.85	–	1.09	2.6	0	−0.4
3	4	4.93	–	0.91	2.1	−0.5	−0.9
4	5	4.99	73±4	0.73	0.3	−2.3	−2.7

Glass II at 450 ± 5°C

Stage	Time, h	Density d, g/cm³	Microhardness H, kg/mm²	ε_g, eV	$\log \sigma_{0e}$	$\log \beta$ As—Se	$\log \beta$ Ge—Ge
1	0	4.591	308±8	1.70	1.7	−0.9	−1.3
2	2	4.60	278±3	1.85	3.1	0.5	0.1
3	3	4.60_8	280±2	1.80	3.2	0.7	0.2
4	4	4.64	–	1.69	3.4_8	0.9	0.5
5	6	4.71	–	1.46	3.7	1.1	0.7
6	8	4.78	123±8	1.24	3.8	1.2	0.8
7	12	4.90	–	0.96	2.7	0.1	−0.3
8	16	4.99	75±4	0.76	1.1	−1.5	−1.9
9	20	5.06	–	0.55	−1.8	−4.8	−4.8
10	27.5	5.07	71±7	0.47	−1.6	−5.0	−4.6

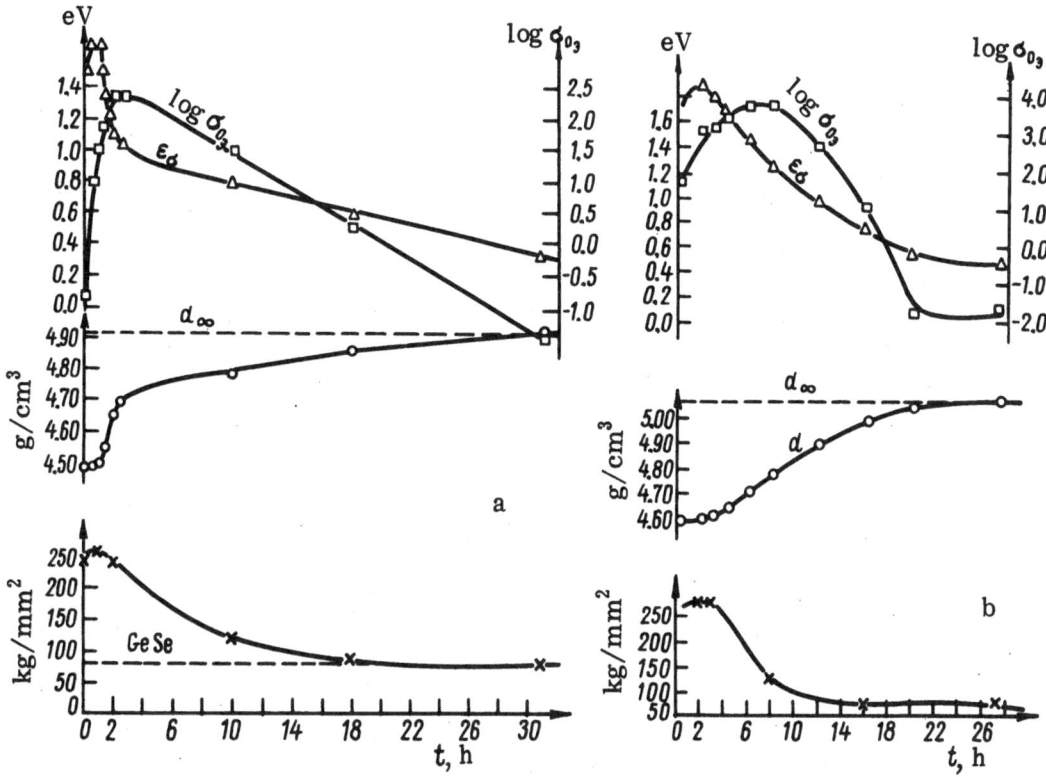

Fig. 4. The rate of change of log σ_{0e}, energy ε_σ, density d, and microhardness H:
a) in Glass I (AsSe$_{5.0}$Ge$_{4.0}$) with long heat treatment at 399°C; b) in Glass II
(AsSe$_{3.0}$Ge$_{2.66}$) with long heat treatment at 400°C.

are stabilized by quenching and produce stresses in the quenched glasses. Such stresses are eliminated, primarily, by the switching of bonds during annealing. In crystallized glasses, as in GeSe polycrystalline aggregates, a decrease in the through conductivity may be the result of the destruction of the continuity of the solid during the coalescence of the crystallites. The data on microhardness in Tables 1 and 2 supports this view.

It follows that neither σ_{0e} nor σ derived from it can be used to measure the progress of the crystallization of the original glass.

It is a different matter with the energy parameter ε_σ. The relative decrease of ε_σ during crystallization is undoubtedly a measure of the accumulation in the crystallizing glass of easily ionized Ge—Ge bonds which in turn characterize the degree of crystallization. It is therefore reasonable to make the assumption that the relative decrease in the energy of ionization of covalent bonds, or in the very closely related conductivity energy, is a measure of the degree of crystallization when the process progresses isothermally in the volume of the glass. Such an assumption is borne out by experience.

6. The use of the relative change in density during the crystallization of glass as a qualitative measure of the degree of volume crystallization is not new. Thus, the degree of crystallization in glass was determined by Slater [40] using the relationship between the percentage of glass which had become crystalline and the density,

$$P_d = \frac{d_t - d_v}{d_{cr} - d_v} \, 100\%,$$

where d_v, d_{cr}, and d_t are respectively, the densities of a material in the vitreous state, the crystalline state, and at a time t during crystallization.

TABLE 3. Kinetic Data for the Crystallization of Glass I

399 ± 4°C

Time, h — Heat treatment	Time, h — Crystallization	% completion of the crystallization process P_d	% completion of the crystallization process P_ε	Rate constant · 10^4, sec^{-1} k_d	Rate constant · 10^4, sec^{-1} k_ε
1.00	0	—	—	—	—
1.25	0.25	—	14	—	1.69
1.50	0.50	11	25	0.67	1.60
2.0	1.0	35	44	1.21	1.58
2.5	1.5	42	48	1.02	0.96
10	9.0	60	64	0.28	0.45
18	17.0	80	82	0.40	0.25
31	30	94	98	0.31	0.36
				0.31± ±0.05	0.36± ±0.07

350 ± 4°C (preheated at 399°C)

Time, h — Heat treatment	Time, h — Crystallization	% completion of the crystallization process P_d	% completion of the crystallization process P_ε	Rate constant · 10^4, sec^{-1} k_d	Rate constant · 10^4, sec^{-1} k_ε
4	0	49	54	—	—
12	8	60	64	0.083	0.083
17	13	65	71	0.076	0.096
23	19	71	75	0.076	0.089
35	31	78	77	0.045	0.064
42	38	80	81	0.070	0.064
47	43	80	82	0.064	0.064
				0.07± ±0.01	0.08± ±0.01

350 ± 4°C (preheated at 450°C)

Time, h — Heat treatment	Time, h — Crystallization	% completion of the crystallization process P_d	% completion of the crystallization process P_ε	Rate constant · 10^4, sec^{-1} k_d	Rate constant · 10^4, sec^{-1} k_ε
—	—	—	—	—	—
3	0	49	68	—	—
5	2	58	71	0.26	0.13
9	6	62	76	0.14	0.13
12	9	62	79	0.10	0.13
14	11	65	81	0.09	0.13
Mean		....		0.11± ±0.02	0.13± ±0.01

From Table 2 it follows that the relaxation of the stress in Glass I (399°C) by annealing was completed at the third stage (the start of the decrease in ε_σ and the ending of the increase in $\log \beta$). Taking $d_v = d_0 = 4.50$ and $d_{cr} = d_\infty = 4.95$ (see Table 1) we obtain

$$P_d = \frac{d_t - d_0}{d_\infty - d_0} 100 = \frac{d_t - 4.50}{0.45} 100\%.$$

In Table 3 we show the results of the corresponding calculations from the data in Table 2.

The crystallization of glasses at 350°C with a preheat at 399° and 450°C, makes it possible to calculate the P_d parameter starting at the second stage (the process must be isothermal).

Analogously, in the case of the crystallization of Glass II (see Table 2) the annealing was completed when the energy reached 1.80 to 1.85 eV. In this case $d_0 = 4.60$ and according to the accepted value of $d_\infty = 5.10$ (Table 1), we obtain

$$P_d = \frac{d_t - 4.60}{0.50} 100\%.$$

The corresponding calculations for Glass II are shown in Table 4.

Assuming now that the relative decrease in the conductivity energy in question can also be used as a measure of crystallization in glass, it is possible that the percentage volume crystallization can be expressed as

$$P_\varepsilon = \frac{\varepsilon_0 - \varepsilon_t}{\varepsilon_0 - \varepsilon_\infty} 100\%,$$

where according to the data in Tables 1 and 2 we have in the case of Glass I: $\varepsilon_0 = 1.70$ eV, $\varepsilon_\infty = 0.30$ eV, and for Glass II: $\varepsilon_0 = 1.80$ eV, $\varepsilon_\infty = 0.40$ eV. It follows then that

$$P_\varepsilon(\text{I}) = \frac{1.70 - \varepsilon_t}{1.40} 100\% \quad \text{and} \quad P_\varepsilon(\text{II}) = \frac{1.80 - \varepsilon_t}{1.40} 100\%.$$

The results of such calculations from the data in Table 2 are shown in Tables 3 and 4.

The results indicate the satisfactory agreement between the values of P_d and P_e, which confirms the validity of utilising both the conductivity energy and the density in determining the degree of crystallization. The unsatisfactory results for Glass I (350°C) with a preliminary heating at 450°C is caused by the appreciable effect of the high temperature heat treatment on the subsequent measurements and we therefore omitted them. In accordance with Rabinovich's data in [33] the closest agreement was observed after 50% of the glass had crystallized.

7. The changes in the density and energy parameters shown in Fig. 4 suggest that the crystallization rate in the glasses is determined by a monomolecular chemical process. At the beginning of this paper we stated that we could assume that the crystallization rate was determined by local activation of the covalent bond. In this case we can use Slater's monomolecular reaction mechanism [40] and the expression for the rate constant becomes

$$k = g\nu \exp\left(-\frac{E}{RT}\right),$$

where $\nu \approx 10^{13}$ sec^{-1} and $g \ll 1$ is the steric factor which expresses the low probability that the appropriate atomic spatial configuration will exist at a given place at the moment when the covalent bond is excited.

The constant k_d and k_ε were calculated by the normal method:

$$k = \frac{2.3}{t} \log \frac{a}{a - x},$$

TABLE 4. Kinetic Data for the Crystallization of Glass II

400 ± 4°C

Time, h (Heat treatment)	Time, h (Crystallization)	P_d	P_ε	k_d	k_ε
2	0	–	–	–	–
8	6	36	44	0.20	0.25
12	10	50	55	0.19	0.21
20	18	64	66	0.16	0.15
32	30	70	71	0.11	0.11
42	40	74	74	0.10	0.08
52	50	76	77	0.08	0.08
		Mean	....	0.11±0.02	0.11±0.03

500 ± 6°C

Time, h (Heat treatment)	Time, h (Crystallization)	P_d	P_ε	k_d	k_ε
2	0	–	–	–	–
3	1	50	54	1.92	2.04
4	2	66	67	1.53	1.47
5	3	78	80	1.41	1.34
		Mean	....	1.47±0.06	1.40±0.06

450 ± 5°C

Time, h (Heat treatment)	Time, h (Crystallization)	P_d	P_ε	k_d	k_ε
3	0	–	–	–	–
4	1	8	8	0.19	0.23
6	3	22	24	0.21	0.26
8	5	36	40	0.24	0.28
12	9	60	60	0.27	0.28
16	13	78	74	0.32	0.29
20	17	92	89	0.41	0.35
27.5	24.5	94	95	0.32	0.34
		Mean	....	0.32±0.04	0.33±0.04

Column headings (all blocks): % completion of the crystallization process (P_d, P_ε); Rate constant ·10^4, sec^{-1} (k_d, k_ε).

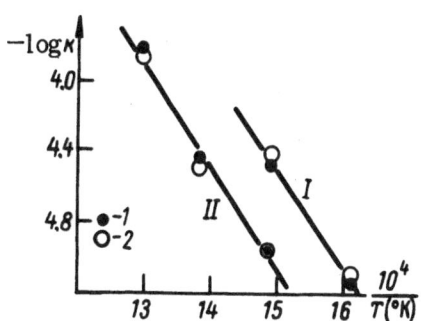

Fig. 5. The temperature dependence $\log k = f(10^4/T)$ in Glass I $(AsSe_{5.0}Ge_{4.0})$ and II $(AsSe_{3.0}Ge_{2.66})$: 1) $\log k_d$; 2) $\log k_\varepsilon$.

where t is the time taken for a quantity of glass x to crystallize in a total initial mass a. In other words, in this expression $a/(a-x)$ corresponds to the reciprocal of the fraction of the initial glass which remains uncrystallized at a time t.

We can therefore express the relationship in question by

$$\frac{a}{a-x} = \frac{d_\infty' - d_0}{d_\infty - d_t} = \frac{\varepsilon_0 - \varepsilon_\infty}{\varepsilon_t - \varepsilon_\infty}.$$

The results of the calculations of the rate constants of crystallization k_d and k_ε based on the data from Table 2 are shown in Tables 3 and 4. These results indicate the wholly satisfactory agreement between the values k_d and k_ε after 50% of the glass has been crystallized.

The temperature dependence of the crystallization rate constants of Glass I and II is given in Fig. 5 and are analytically expressed in the form

$$\log k_I = -\frac{5740}{T} + 4.08,$$

$$\log k_{II} = -\frac{5800}{T} + 3.60,$$

which corresponds to the value of the activation energy E = 1.15 eV (26.3 ± 0.2 kcal/mole) and to the values $\nu\, g_I = 1.2 \cdot 10^4$ and $\nu\, g_{II} = 4.0 \cdot 10^3$, from which we derive the approximate values $g_I \approx 1 \cdot 10^{-9}$ and $g_{II} \approx 4 \cdot 10^{-10}$. The inequality $g_{II} < g_I$ is perhaps due to the nonstoichiometry of composition No. II.

We cannot exclude the possibility that the increased value of the constants when P < 50% (see Tables 3 and 4) relates mainly to the process of the generation of crystallization centers, while the lower value of the constants when P > 50% relates to the crystal growth process. However, keeping to strict consideration, we suggest that it is premature to attempt to define the crystallization rate process mechanism more exactly on the basis of quite limited experimental material. It is possible that the superposition of both processes rules out the possibility of separating them completely. Such a subdivision of the process is even more impractical since we can consider that both components of the crystallization process in glasses are ultimately determined by local structural chemical reorganizations with an activation energy required for the switching (transformation) of the covalent bond. From this point of view it would be interesting to determine the activation energy of the viscous flow in glasses of the compositions considered here and to compare these values with those obtained in this work.

References

1. G. Tammann, Kristallisieren und Schmelzen, Leipzig (1903); Aggregatzustande, Leipzig (1923); Glaszustand, Leipzig (1933).
2. V. D. Kuznetsov, Crystals and Crystallization, Gostekhizdat, Moscow (1953).
3. W. Eitel, The Physical Chemistry of the Silicates, Chicago (1954).
4. M. Volmer, Z. phys. Chem., A102:267 (1922); Kinetik der Phasenbildung, Dresden (1939).
5. Growth of Crystals (A. U. Shubnikov ed.), Vols. 1-3, Consultants Bureau, New York (1958-1962).
6. A. A. Leontjewa, Acta Physicochim. URSS, 13:423 (1940); 14:245 (1941); J. G. Morley, Brit. J. Appl. Phys., 12:10 (1961).

7. I. N. Stranskii, and Katishev, Usp. Fiz. Nauk, 21:408 (1939); B. Honigman, Gleichgewichts und Wachstumformen von Kristallen, Darmstadt (1958).

8. W. J. Dunning, Chemistry of the Solid State, London (1955).

9. R. Collongues, Bull. Soc. Chim. Fr., 265 (1957).

10. The Mechanical Properties and the Structure of Inorganic Glasses (Material for the symposium in Leningrad April 10-12, 1962), Izd. Akad. Nauk SSSR, Leningrad (1962).

11. Proceedings of the Seminar on the Theory of Crystallization of Inorganic Glasses in Leningrad April 12-14, Izd. Akad. Nauk SSSR, Leningrad (1962).

12. ● R. L. Myuller, Izv. Akad. Nauk SSSR, 4:607 (1940), ● this volume, p. 170; in: The Structure of Glass, Vol. 2, Consultants Bureau, New York (1960), p. 50.

13. Kuan Han Sun, J. Amer. Cer. Soc., 30:277 (1947).

14. F. E. Simon, Ergeb. exakt. Naturwiss, 9:244 (1930).

15. R. O. Davies and G. O. Jones, Advances in Physics, 2(7):370 (1953).

16. R. L. Myuller, Zh. Prikl. Khim., 28:363 and 1077 (1955).

17. O. V. Mazurin (editor), The Structure of Glass, Vol. 4, Consultants Bureau, New York (1966).

18. O. V. Mazurin and V. A. Tsekhomskii, Leningr. Tekhnol. Inst., 59:33 and 36 (1961).

19. R. L. Myuller, Vest. Leningr. Gos. Univ., No. 22, no. 4, 86 (1961); in: Physics, Proceedings of the 20th Scientific Conference of the LISI, Leningrad (1962), p. 18.

20. R. L. Myuller, Zh. Tekhn. Fiz., 25:1567 (1955), ● this volume, p. 43.

21. P. T. Kozyrev, Zh. Tekhn. Fiz., 26:255 (1956); Abstract of Candidate's Dissertation, Izd. Akad. Nauk SSSR, Leningrad (1958).

22. L. A. Baidakov, Z. U. Borisova, and R. L. Myuller, Zh. Prikl. Khim., 34:2446 (1961), ● this volume, p. 133.

23. Z. U. Borisova, E. V. Shkol'nikov, and I. I. Kozhina, Vestn. Leningr. Gos. Univ., No. 22, Iss. 4, p. 114 (1962).

24. A. A. Lebedev, Tr. Gos. Optich. Inst., 5:1 (1920).

25. A. A. Kudryavtsev, The Chemistry and Technology of Selenium and Tellurium, Izd. Vysshaya Shkola, Moscow (1961).

26. S. A. Shchukarev and R. L. Myuller, Zh. Fiz. Khim., 1:625 (1930).

27. N. G. Alekseev, V. N. Prokhorov, and K. V. Chmutov, Electronic Instruments and Circuits for Physical Chemistry, Goskhimizdat, Moscow (1961).

28. R. L. Myuller, Tomsk. Univers., 145:33 (1957).

29. I. D. Tykachinskii and E. S. Sorkin, Proceeding of the Seminar on the Theory of the Crystallization of Inorganic Glasses in Leningrad, April 12-14, 1962, Izd. Akad. Nauk SSSR Leningrad (1962).

30. Ya. I. Frenkel', J. Chem. Phys., 7:200 and 538 (1939).

31. B. Ya. Pines, Zh. Éksp. Teor. Fiz., 18:29 (1948).

32. F. K. Gorskii, Zh. Éksp. Teor. Fiz., 18:45 (1948).

33. É. M. Rabinovich, Dokl. Akad. Nauk SSSR, 138:159 (1961).

34. A. S. Pashinkin, Liu Tsun-hua, Ya. M. Nesterov, and A. V. Novoselova, Proceedings of the All-Union Conference on Semiconducting Compounds, Leningrad, December 18-23, 1961, Izd. Akad. Nauk SSSR, Moscow—Leningrad (1961), p. 52.

35. Z. U. Borisova, R. L. Myuller, Chin Ch'eng-ts'ai, Zh. Prikl. Khim., 35:774 (1962); R. L. Myuller, L. A. Baidakov, and Z. U. Borisova, Vestn. Leningr. Gos. Univ. No. 10, Iss. 2, p. 94 (1962), ● this volume, p. 143.

36. H. Krebs, Z. anorg. allg. Chem., 278:82 (1955).

37. E. Mooser and W. B. Pearson, J. Electronics, 1:629 (1956).

38. L. A. Baidakov, Z. U. Borisova, and R. L. Myuller, Zh. Prikl. Khim., 33:2446 (1961).

39. A. V. Danilov and R. L. Myuller, Zh. Prikl. Khim., 35:2012 (1962).

40. N. B. Slater, Proc. Royal Soc., A194:112, 1948; Theory of Unimolecular Reactions, Ithaca, New York (1959).

Bond Energy, Ionization, and the Conductivity Modulus of Vitreous Semiconductors as Functions of Composition*

1. A vitreous solid does not have a long range geometric order but it may have a particularly interesting long range structural chemical order. The energetics of short range order, defined by the covalent character of the chemical bonds between the atoms, stand out in the vitreous state. From a chemical point of view, the chalcogenide glasses in the germanium-arsenic-selenium system are worth considering further, as covalent heterobonds are expressed most clearly in this sytem. To a first approximation we can consider the value of the band gap, which is related to the intrinsic conductivity of glass, as a measure of the difference between the energy of the electron-pair and the single-electron chemical bonds between the respective atoms, or alternatively as the ionization energy of a covalent bond. It is clear, then, that the value of the band gap must be less than the value of the energy required to break the corresponding chemical bond.

An examination of the activated conversion of a two-electron to a one-electron bond has established that the pre-exponential term σ_0 is proportional to the concentration of covalent bonds belonging to the majority type of structural units in the glass. The logarithm of the proportionality constant, i.e., the logarithm of the ratio σ_0 to the concentration ν of the covalent bonds, was called the conductivity modulus [1]. In the glasses we examined the theoretical value of the conductivity modulus was 4.6, within one unit. The difference between the experimental and the theoretical value of the modulus is given by the parameter

$$\log \beta = \log (\sigma_0/v) \exp (-4.6).$$

A study of the simplest binary vitreous selenide systems has established that the experimental and theoretical values of the conductivity modulus are in agreement and thus $\log \beta$ is zero within one unit for three-dimensional, covalently bonded chemical structures which allow through conductivity.

We have very recently studied complex multicomponent vitreous systems and made a more detailed study of the Ge−As−Se and Ge−As−S systems. The experimental values of the conductivity modulus for annealed glasses, in all cases with continuous chemical structures, corresponded to the theoretically calculated values. We must further consider the comparison of the values of the chemical bond energy with the values of the band gap.

2. At the present time completely reliable data are not available for bond energy. The values derived from the calculations of atomization energy, with the addition of the critically verified data in [2], deserve our attention. In this case the following values for the energy of unit homobonds in kcal/mole are obtained:

$$D(\text{S} - \text{S}) = 65; \quad D(\text{As} - \text{As}) = 46; \quad D(\text{Se} - \text{Se}) = 49; \quad D(\text{Ge} - \text{Ge}) = 46.$$

*R. L. Myuller, Izv. Akad. Nauk SSSR, ser. fiz., 28:1279 (1964).

Subsequently, the breaking energy of the heterobonds was calculated from thermochemical data and the electronegativity data of Pauling, giving

$$D(As - S) = 61; \quad D(Ge - S) = 68;$$
$$D(As - Se) = 52; \quad D(Ge - Se) = 56;$$
$$D(Ge - As) = 47.$$

As a result of these thermochemical calculations, we established a series of structural chemical reorganizational reactions occuring with the gradual increase in the arsenic and germanium concentrations in sulphide and selenide glasses ($-\Delta H_0$ at 298°K, in kcal):

$$3S_2 + 2As_2 \rightarrow 4AsS_{1.5} \quad 66 \, (A);$$
$$2S_2 + Ge_2 \rightarrow 2GeS_2 \quad 100 \, (B);$$
$$2Se_2 + 2As_2 \rightarrow AsSe_{1.5} \quad 54 \, (A);$$
$$2Se_2 + 2Ge_2 \rightarrow 2GeSe_2 \quad 68 \, (B);$$
$$4AsS_{1.5} + Ge_2 \rightarrow 2As_2S_2 + 2GeS \quad 28 \, (C);$$
$$2AsS_{1.5} + GeS \rightarrow As_2S_2 + GeS_2 \quad 14 \, (C);$$
$$As_2S_2 + Ge_2 \rightarrow As_2 + 2GeS \quad 28 \, (D);$$
$$As_2S_2 + 2GeS \rightarrow As_2 + 2GeS_2 \quad 28 \, (D);$$
$$4AsSe_{1.5} + Ge_2 \rightarrow 2As_2Se_2 + 2GeSe \quad 16 \, (C);$$
$$2AsSe_{1.5} + GeSe \rightarrow As_2Se_2 + 2GeSe_2 \quad 8 \, (C);$$
$$As_2Se_2 + 2GeSe \rightarrow As_2 + 2GeSe_2 \quad 16 \, (D).$$

We must note here that these calculations are based on the current standard enthalpy values and these unfortunately do not allow us to calculate the changes in the free energy in the melt and on subsequent annealing. The quite appreciable enthalpy changes (50 to 100 kcal) do not leave any doubt about the occurrence of these reactions leading to the formation of the following structural units; the sulfides $AsS_{1.5}$, GeS_2 ; and the selenides $AsSe_{1.5}$ and $GeSe_{2.0}$. This case is not quite so good for the subsequent reactions producing the lower selenides, As_2Se_2, and GeSe; and sulfides As_2S_2, and GeS, with an enthalpy change of 10–30 kcal. In these cases we can only suggest that there is a high probability of such reactions occurring. The subsequent analysis of the semiconducting and mechanical properties of the chalcogenides in question has so far confirmed the sequence, during glass formation, of the chemical processes which we have noted here.

Both in selenides and sulfides processes A and B occur first, followed by process D. These structural chemical reorganizations are accompanied by a marked increase in the packing density of the atoms (by 3–10%), and by the establishment of stable and reproducible through conductivity ($\log \beta \approx 0$). Glasses of intermediate compositions which correspond to the occurrence of process C in the melt, with the formation of As_2Se_2 and As_2S_2 structural units, are characterized by a looser atomic packing (by 3–6%), by the instability of the conductivity, and by its extreme sensitivity to the temperature-time schedule of the glasses during preparation.

The accumulated experimental material allows us to compare the values of the band gap with the calculated values of bond energy.

As can be seen from Table 1, the energy of breaking of bonds is invariably greater than the width of the band gap. As we have already noted, the width of the band gap corresponds to the escape of only one electron from the electron-pair bond with the retention of a one-electron bond. We should note further the closeness of the values of the band gap to those of the optical absorption edge. Therefore, the carrier mobility activation energy is small. We must note also that the activation energy for the escape of atoms from the glass into solution is less than the energy of breaking of the bond and of the band gap.

3. We noted above that the addition of germanium to arsenic selenide and sulfide may be accompanied by the formation of As_2Se_2 and As_2S_2 structural units which loosen the glass and

TABLE 1

Unit bond	Bond energy	Optical absorption edge	Band gap	Mobility activation energy	Activation energy of dissolution
	eV				
As—Se	2.3	1.7	1.7	≤0.05	0.8
Ge—Se	2.4	1.8	2.2	~0.2	0.7
As—S	2.7	2.2	2.2	≤0.05	–

have a fundamental effect on the conductivity. The systematic study by Borisova and her colleagues of the effect of adding impurities from the Group II-VI elements to arsenic selenide confirmed, on the whole, the previous observations by Kolomiets et al. [3], that small amounts of impurity did not have any effect on the conductivity of glasses. If, however, the impurity markedly alters the structural chemistry of the glass then the effect on conductivity becomes apparent. An interesting case of a significant effect due to an added component is found when we introduce copper into $AsSe_{1.5}$ arsenic selenide. We studied this experimentally in [4] and consider it worthwhile to examine the principles involved. With successive introductions of copper up to 19 at. %, we observed a significant change in the band gap which decreased monotonically from 1.8 to 0.9 eV. The conductivity modulus remained constant and was within 0.1 of the theoretical value. In this case the monotonic character of the change in the width of the band gap at quite low current carrier mobilities is important. We established an interesting empirical quantitative law for the decrease in the value of the band gap. We observed that the product of the band gap energy and the fourth root of the atomic concentration of copper in the glass was constant. It is remarkable, in that we had previously established such a law for sodium ion conductivity in silicate glasses [5].

Structural chemical reorganizations in glass occur, not only with the introduction of new components but also during the crystallization of the glass. Thus, x-ray data show that a GeSe crystalline phase [6], most probably in the form of crystallites highly dispersed in the volume of the glass, develops on volume crystallization of some arsenic—germanium—selenide glasses. With such a structural change the monotonic contraction of the band gap from 1.70 to 0.40 eV was observed [7]. It was established that the proportion of the glass which had crystallized, determined by Rumsh et al. [8] by x-ray studies and by us from the density change, is numerically equal to the relative change in the band gap. In other words, the band gap of the partially crystallized glass is an additive function of the value of the band gap of the particular substance in the vitreous and crystalline states.

The monotonic change in the band gap during structural transformations of glass which we have established experimentally in a number of cases is difficult to explain at present by a quantitative theory.

References

1. R. L. Myuller, Zh. Prikl. Khim., 35:541 (1962), ● this volume, p. 123.
2. V. N. Kondrat'eva (editor), The Breaking Energy of Chemical Bonds, Ionization Potentials and Electron Affinity, Izd. Akad. Nauk SSSR, Moscow (1962).
3. B. T. Kolomiets and T. F. Nazarova, Fiz. Tverd. Tela, 2:174 (1962).
4. A. V. Danilov and R. L. Myuller, Zh. Prikl. Khim., 35:2012 (1962).
5. R. L. Myuller, Fiz. Tverd. Tela, 2:1333 (1960).
6. Z. U. Borisova, E. V. Shkol'nikov, and I. I. Kozhina, Vestn. Leningr. Gos. Univ., Vol. 17, No. 22, Iss. 4, p. 114 (1962).

7.   R. L. Myuller and E. V. Shkol'nikov, Vestn. Leningr. Gos. Univ., Vol. 17, No. 22, Iss. 4, p. 119 (1962), ● this volume, p. 150.
8.   E. V. Shkol'nikov, M. A. Rumsh and R. L. Myuller, Fiz. Tverd. Tela, 6:796 (1964).

PART 3

THE STRUCTURE OF VITREOUS MATERIALS

The Problem of the Vitreous State*

The addition of a basic oxide which reacts with it to an anhydride is accompanied by the break-up of the network structure, by a decrease in the number of covalent bonds in the system, and by an increase in electrostatic and dipole interactions. Accordingly, as the alkali content is increased, the viscosity of the system in the molten state decreases and the tendency of the system to crystallize increases. Three cases are possible:

1. The resultant salt-like polar groups remain connected to the remaining anhydride network and are randomly distributed throughout it. This system is homogeneous.

2. The salt-like groups break their chemical bonds with the anhydride network and are randomly distributed in the form of molecular, dispersed inclusions. Again, we have a homogeneous system.

3. The salt-like polar molecules which are not connected to the anhydride network by chemical forces interact with each other and form associated complexes, the number and size of which increase with the increase in the basic oxide content of the glass. In this case, we have a complicated microheterogeneous system.

On analysing the minimum in the electrical conductivity observed in glasses, we came to the conclusion, as early as 1934, that alkali borate glasses have a microheterogeneous structure. New data from the conductivity measurements on ternary systems points even more specifically to such a structure of alkali glasses.

The significant concentration dependence of the specific conductivity of simple lithium and potassium glasses had been noted earlier. The experimental results of Markin indicate that this type of dependence is retained in complex lithium—potassium glasses, suggesting that the character of the interaction of similar salt-like particles is the same in complex as in the simple lithium—potassium glasses.

The latter results may be interpreted if we assume that in alkali glasses there is a marked clustering of similar salt-like particles in the form of ultramicrocrystalline aggregates or of some associated complexes.

The small deviation from additivity in the conductivity of sodium potassium borate glasses observed in the data of Markin, Brodskaya, and Tatarinova is clearly the result of the well known mixing of the microcrystals of the salt-like components containing cations of similar radii.

The experimentally established fact that the displacement energy for cation conductivity is the same in both complex and simple glasses supports the interpretation given here. The previously established minimum in the molar conductivity, the increase on quenching in the conductivity of the glass, and the effect of divalent cations, are all in conformity with this model.

An analysis of the conductivity of silicate glasses indicates that they are analogous in nature to borate glasses.

*R. L. Myuller, Byull. Vsesoyuz. Khim. Obshchestva im. D. I. Mendeleeva, No. 6, pp. 12-13 (1939).

Conductivity Data and The Structure of Solid Glasses*

1. In using the term "the vitreous state" to designate the state in which glass exists, we certainly wish to underline the distinction between this state and the liquid or crystalline states. There has recently been an attempt, by all those research workers involved in the study of crystalline bodies, to reveal the element of disorder in the particle distribution in them. The opposite can also be observed in the study of the liquid state: the attempt to reveal the element of order in the particle distribution. We can agree with Frenkel' that before discussing the structure of one or other system we must consider the structure of that unit on which the model is based. It is not, however, sufficient to confine ourselves to three-dimensional geometrical models.

In contemporary crystal chemistry, in addition to the analysis of these geometrical relationships, some attention has been devoted to the problem of the nature of the chemical bonds which are responsible for the strength of substances in the crystalline state. As a worker in the field of the liquid and solid state I believe that insufficient attention has been paid to this aspect of the problem. Until this is rectified it is not so important, for example, to know whether there are or are not some sort of crystalline inclusions in glass. A knowledge of the fundamental structural unit and the nature of its chemical bonds is the more essential as far as problems of the vitreous states are concerned.

Essentially, the very tendency of particular substances to crystallize is, in my opinion, associated with the predominance in them of particular directional bonds with a fairly small radius of action. These are primarily powerful covalent bonds acting between the atoms of high melting oxides of the Groups III, IV, and V elements of the Periodic System. In these cases the valencies of the elements are significant, leading as they do to trigonal and tetrahedral orientation of the chemical bonds which ensure, in the oxygen compounds, the characteristic three-dimensional crystalline lattices and the amorphous atomic networks, as Zachariasen and Warren demonstrated so well.

The covalent bonds in an "atomic" network (for example quartz) produce, at not too high temperatures, atomic oscillations of smaller amplitude than the ionic oscillations in an ionic lattice (for example in sodium chloride). This difference, I believe, sets in relief the reason for the high viscosity and increased activation energy of atomic rearrangement observed in substances with a tendency to crystallize.

The study of organic vitreous substances at low temperatures provides us with strong arguments in favor of the view that the particles in them are linked by directional van der Waals forces.

It is doubtful, however, whether a silica melt can consist essentially of molecules at temperatures slightly above the melting temperature. This is the more doubtful in that weak intermolecular van der Waals' forces cannot provide at a high temperature sufficient stability in the

* R. L. Myuller, Izv. Akad. Nauk SSSR, ser. fiz., 4:607 (1940).

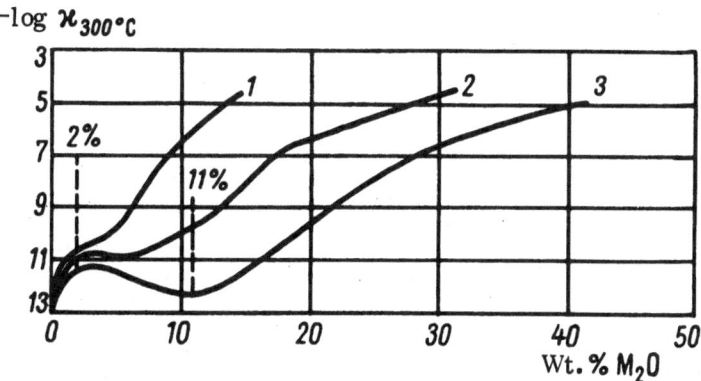

Fig. 1. Dependence of the log specific conductivity at
300°C of alkali borate glasses on the alkali oxide content expressed in wt.%: 1-3) glasses containing various
oxides.

liquid silica. High melting and relatively involatile silica, boric anhydride and similar oxides
are, from their structure in the melted state, atomically constructed systems and their strength
is the result of powerful covalent bonds.

It is thus almost impossible to produce a complete analogy between volatile organic and
high-melting inorganic glasses.

The comments below can be related only to vitreous systems of chemical composition similar to normal high melting glasses based on an anhydride like silica, boric anhydride, or phos-
phoric anhydride, etc.

Above all, we shall attempt to examine high melting oxygen containing vitreous systems
from the point of view of the presence in them of two types of structural units: covalently bonded atoms and electrostatically interacting ions. It will then be shown that such vitreous systems are not homogeneous but that there is some differentiation of the components in them.

As early as 1935 we expressed the opinion [1] that in those vitreous systems with a significant concentration of alkali oxides an association of polar groups occurs. With the existence
now of experimental data for the conductivity of three-component systems we can be more positive in the expression of this opinion. We can consider that the component parts of a system,
uniform in their composition and in the character of their chemical bonds, are present in differentiated groups which thus give the glass its nonuniform character.

In this connection the nature of the chemical bond is very important in the electrochemistry of liquid systems, and consequently, in the electrochemistry of solidified vitreous systems.

As a result of many years study of the conductivity of glasses we have collected a wealth
of experimental material. Limiting ourselves to presenting only the data relating to the electrochemistry of alkali borate glasses we shall try to extend the concepts outlined above.

2. Above all, we shall be concerned with the experimental dependence of the log specific
conductivity on the weight per cent of alkali oxides.

We notice the extremely large rise in conductivity (in potassium glasses by 10^7) at a certain concentration. This rise begins (Fig. 1) in glass containing different oxides at different
weights per cent of the oxide. It is not difficult to see that this is the result of an unfortunate
choice of the units for measuring the alkali content in the glasses. In fact, when we analyze the
data for specific conductivity we are forced to realize that the latter is proportional to the volume

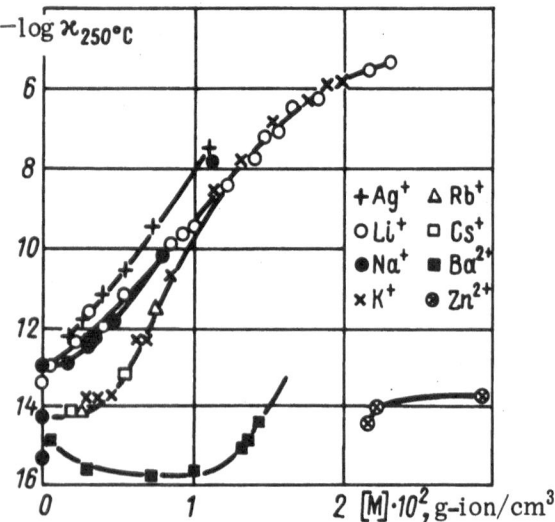

Fig. 2. Dependence of the log specific conductivity at 250°C of various glasses on their concentration of metal oxides expressed in g-ions of metal per cm³.

concentration of alkali ions. It is therefore advisable to plot the concentration as the volume concentration expressed in gram-ions of alkali metal per cubic centimeter [M], rather than as weight per cent.

By making this change we get the new curves shown in Fig. 2. We see here that for high alkali concentrations a curve is obtained which is common to all forms of alkalis. At low concentrations there is a fan-like divergence of the curves for glasses containing different alkali oxides.

We should note that the conductivity was measured over the temperature range 200-300°C. This range corresponds to the critical temperature region in glasses low in alkali but lies considerably below that region in glasses which are rich in alkali. Thus, low alkali glasses were studied under conditions where structural changes were occurring, the rate of which was commensurate with the rate of measuring. We studied only alkali-rich glasses at fairly low temperatures when no structural changes were observed throughout the period of the conductivity measurements. As a result, the temperature coefficient of the conductivity of these glasses was free from the effect of the structural changes observed at higher temperatures.

We are confining ourselves here to an examination of the data for the conductivity of alkali-rich glasses. The conductivity of such glasses agrees satisfactorily with our theory of electrolytic dissociation in high melting oxygen-containing glasses. This theory points to the related character of the ionic electrical conductivity of glasses and the conductivity of ionic crystals to which, as is well known, the Frenkel'-Jost-Schottky theory is applicable [3].

There are, according to our proposals, two types of cation sites: a position near ionized oxygen atoms $(R-O^-Na^+)$, and a position near non-ionized oxygen atoms $(R-O-R)Na^+$. In the first case, the metal ion is connected to the oxygen by electrostatic coulomb forces and being relatively strongly attached can be considered as bonded. The $O^- -Na^+$ group can here be considered as "electrolytically undissociated." With the dissociation of such a group the alkali metal ion can move, as a result of thermal motion, inside the atomic glass network, transferring by jumps from one nonionized oxygen atom to another and being connected to them by polarization forces. There will be a particular equilibrium distribution of alkali cations between ionized and non-ionized oxygen atoms of the glass corresponding to a given temperature. In this case the lack of thermodynamic equilibrium of the glass as a whole does not prevent the examination of the equilibrium distribution of cations, since the structural changes were excluded by studying alkali-rich glasses. The equilibrium distribution of cations may be expressed as a chemical equation

$$\overset{|}{\underset{|}{O}}{}^-Na^+ + \overset{|}{\underset{|}{O}} \rightleftarrows \overset{|}{\underset{|}{O}}{}^- + Na^+\overset{|}{\underset{|}{O}}$$

We have produced a theoretical expression for the molar conductivity of glasses $\lambda = \varkappa / [M]$ which after rearrangement becomes

$$\lambda = \frac{\varkappa}{[M]} = 6700 z^2 \sqrt{\gamma z} \cdot e^{-\frac{E}{2RT}}, \tag{1}$$

where $\varkappa$ is the specific conductivity; z is the valency of the moving cations; E is the dissociation energy; R is the gas constant; T is the absolute temperature; and finally γ is the ratio of the number of un-ionized to ionized oxygen atoms:

$$\gamma = \frac{[O]}{[O^-]}.$$

The familiar empirical linear dependence of the log specific conductivity on the reciprocal of the temperature:

$$\ln \varkappa = -\frac{A}{T} + B,$$

can without difficulty lead to the empirical expression for molar conductivity:

$$\lambda = 10^{0.4343B - \log [M]} \cdot e^{-\frac{A}{T}}. \tag{2}$$

TABLE 1. Borate Glasses

Glass No.	$[M] \cdot 10^2$, g-ion/cm^3	0.4343 B-log [M] (exp. determined [2])	log $(6700 z^2 \sqrt{\gamma z})$ (theoretically calculated)
System Li$_2$O—B$_2$O$_3$			
111	0.84	4.0	4.3
17	0.92	4.9	4.3
112	1.00	4.2	4.3
113	1.22	3.6	4.2
114	1.40	3.6	4.2
18	1.47	4.6	4.2
115	1.55	3.8	4.2
116	1.67	3.9	4.1
30	2.15	3.9	4.1
System Na$_2$O—B$_2$O$_3$			
7	0.45	4.7	4.4
8	0.79	3.5	4.3
9	1.10	3.8	4.2
10	1.80	3.0	4.1
11	2.3	3.1	4.0
System K$_2$O—B$_2$O$_3$			
52	0.84	3.9	4.3
53	1.13	3.4	4.2
54	1.31	3.7	4.1
55	1.52	3.6	4.1
56	1.76	3.3	4.0
57	1.88	3.3	4.0
58	1.98	3.4	4.0
Mean 3.7			4.2
System Ag$_2$O—B$_2$O$_3$			
148	0.54	3.8	4.4
149	0.72	3.4	4.3
150	1.10	3.3	4.2
Mean 3.5			4.3

TABLE 2. Sodium Silicate Glasses

Glass No.	$[M] \cdot 10^2$, g-ion/cm^3	$0.4343\,B - \log[M]$ (Exp. determined [5])	$\log(6700\,z^2\,\sqrt{\gamma z})$ (theoretically calculated)
114	4.2	3.6	3.7
115	3.8	3.8	3.8
116	3.3	4.1	3.8
117	2.9	3.7	3.9
118	2.7	3.7	3.9
119	2.4	3.6	4.0
120	2.2	3.6	4.0
121	2.13	3.7	4.0
122	2.05	3.5	4.0
123	1.82	3.4	4.1
124	1.52	3.5	4.1
126	1.18	3.6	4.2
127	0.90	3.4	4.2
128	0.63	3.4	4.3
Mean		3.6	4.0

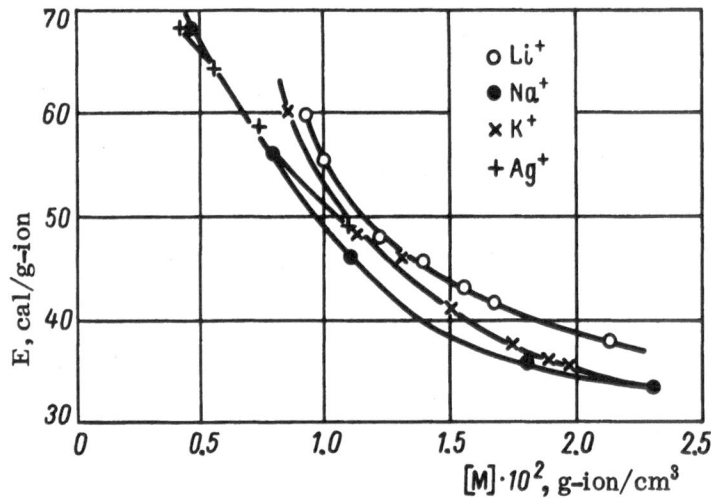

Fig. 3. Dependence of the dissociation energy of the polar groups on the concentration of univalent metal oxides in the glass, expressed in g-ions of metal per cm^3.

By comparing the empirical expression, Eq. (2) with the theoretical, Eq. (1) it is easy to obtain, by equating $A = E/2R$

$$0.4343B - \log[M] = \log\left(6700z^2\,\sqrt{\gamma z}\right). \tag{3}$$

A comparison between the empirical and theoretical expressions in the logarithmic form was used, since the calculation of the theoretical value is only accurate to an order of magnitude.

The calculated values of both sides of Eq. (3) are shown in Tables 1 and 2.

It is evident from Tables 1 and 2 that the mean value of the empirical expression in question for alkali borate glasses is 3.7, while the theoretical value is 4.2. In the case of silver borate glasses the respective values are 3.5 and 4.3.

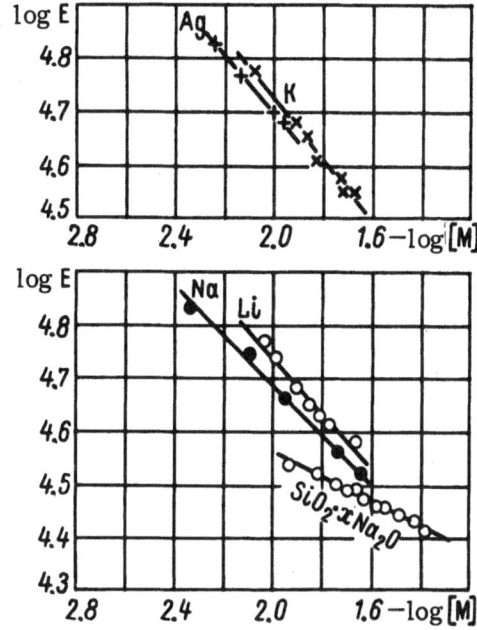

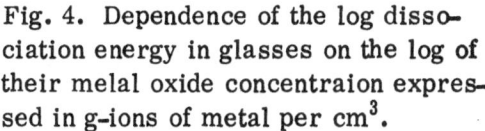

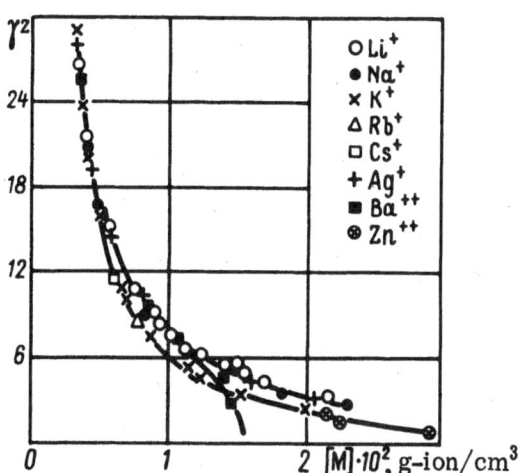

Fig. 4. Dependence of the log dissociation energy in glasses on the log of
their melal oxide concentraion expressed in g-ions of metal per cm^3.

Fig. 5. The dependence of the ratio of
nonionized to ionized oxygen atoms on
the concentration of metal oxides in the
glass, expressed in g-ions of metal per
cm^3.

Using the data obtained by Turner, Seddon, and Tippett [5] for silicate glasses we obtain
the respective mean values of 3.6 and 4.0.

Such good agreement indicates that our proposed theory of electrolytic dissociation is applicable to high melting oxygen-containing inorganic silicate and borate glasses. The empirical
coefficient A can be related to the dissociation energy E noted above:

$$A = \frac{E}{2R} \qquad (4)$$

(according to Eqs. (1) and (2)).

The values of the dissociation energy (in kcal/g-ion) derived thus, for various glasses
with a different alkali content, are given in Fig. 3. In all the glasses an abrupt decrease is observed in the dissociation energy with an increase in the alkali content. The dissociation energy
of a cation in borax in the vitreous state is approximately 35 kcal/g-ion. According to Frenkel's
theory it should be close to 40 kcal/g-ion.

In 1932 we noted [4] the empirical dependence of log E on log M (Fig. 4). The dependence
of E−[M] (Fig. 3) is incomparably more interesting. In order to understand this we must con-
cern ourselves with the characteristic decrease in the dissociation energy with an increase in
the metal ion concentration which is accompanied by a simultaneous increase in the concentration of ionized oxygen atoms. Under these conditions, as a result of the proximity of the electrostatically interacting ionized atoms in the glass, the metal ions on dissociation escape less
and less from the field of influence of the electrostatic forces. This, clearly, is responsible for
the decrease in the dissociation energy. The change in the relationship between the number of
sites containing the metal ions in an undissociated state near ionized oxygen atoms and in a

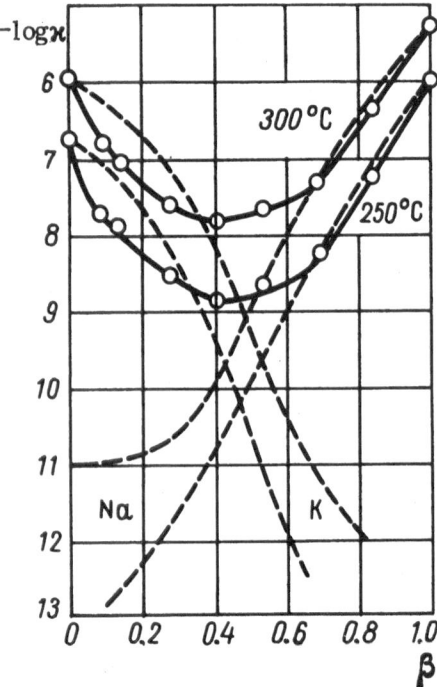

Fig. 6. The dependence of the log specific conductivity of alkali borate glasses on the ratio between the Na and Na+K concentration expressed in g-ions per cm³.

dissociated state near non-ionized oxygen atoms is essential in this case. In other words, the decisive role in dissociation must be played by the ratio of the number of ionized to nonionized oxygen atoms.

Let us consider the empirical dependence of the ratio between nonionized and ionized oxygen atoms on the cation concentration (Fig. 5). We must note here the hyperbolic character of the dependence of the expression γz on [M], reminiscent of the dependence of the dissociation energy on cation concentration (see Fig. 3). By comparing the E and γ parameters we established that they were linearly related. In fact, the dissociation energy is given by the relationship $E = E_0 + a\gamma$, where a is a constant.

We can then deduce that the electrolytic dissociation energy in glasses, like the molar conductivity, is determined by the fundamental ratio of the ionized to the non-ionized oxygen atoms in the glass.

3. In order to clarify the problem of nonuniformity in glasses let us consider the results of studies involving the specific conductivity of three-component alkali borate glasses.

An analysis of the conductivity data for $Na_2O-K_2O-B_2O_3$ systems is shown in Fig. 6. The values of the log specific conductivity is plotted against a parameter given by the ratio

$$\beta = \frac{[Na]}{[Na] + [K]} \cdot$$

The average total concentration [Na] + [K] = [M] was $1.7 \cdot 10^{-2}$ g-ion/cm³. The continuous curves represent the conductivity of the three-component systems in question; the dotted curves represent the conductivity of purely sodium and purely potassium glasses with concentrations [Na] = $1.7 \cdot 10^{-2} \cdot \beta$ and [K] = $1.7 \cdot 10^{-2} (1 - \beta)$, respectively. The upper curve refers to measurements at 300°C, the lower to those at 250°C.

The closeness of the curves corresponding to the three-component and two-component systems in Fig. 6 indicates the generality of the nature of the change in the conductivity of complex and simple glasses. Such a quantitative agreement between the conductivity values is the result of the closeness in the values of the dissociation energy in these three-component glasses and in the corresponding potassium and sodium two-component glasses (Fig. 7).

The dissociation energy in the case of the two-component glasses is a function of the alkali content and thus depends unequivocally on their structure. The retention in the three-component glasses of this dependence of the dissociation energy on the concentration of each type of ion separately points clearly to the preservation in the three-component glasses of the structures which are peculiar to the corresponding simple two-component glasses.

As the change in dissociation energy is undoubtedly the result of the interaction of the ionic constituents we must assume that in three-component glasses the same type of interaction is retained as is observed in two-component glasses. We can therefore conclude that there is an interaction of ionic groups of one type in the glass. The quantitative agreement noted earlier

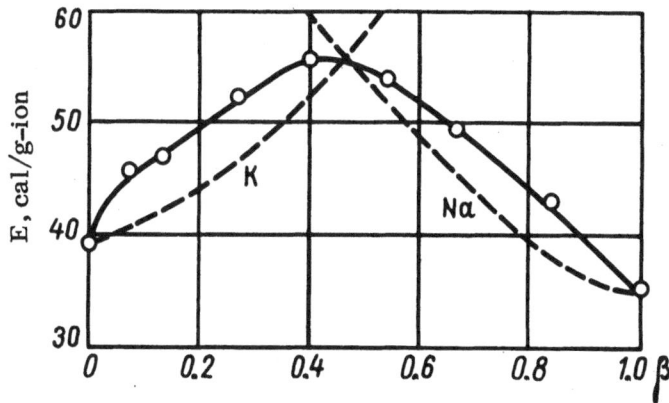

Fig. 7. The dependence of the dissociation energy of three-component (continuous curve) and two-component (broken curve) alkali borate glasses on the ratio between the Na and Na+K concentration expressed in g-ions per cm^3.

indicates that there is weak interaction of groups of different types. If the opposite were the case we would have observed significant discrepancies, which would be the greater the more the overall alkali concentration remained unchanged throughout the whole system.

Thus, the presence in these glasses of differentiated groups each containing the same type of components follows from the above. In other words we have here a real indication of the nonuniform structure of glass with respect to its composition.

There is no opportunity here to dwell in more detail on the results of our research but we shall note only that the data for the vitreous systems, $Li_2O-Na_2O-B_2O_3$, $Na_2O-K_2O-B_2O_3$, and $Li_2O-K_2O-B_2O_3$, confirm the above conclusions and a strict quantitative analysis of them proves the compositional inhomogeneity of these glasses.

To summarize briefly we can note the applicability to borate and silicate glasses of the covalent-ionic structural concepts which provide a quantitative theory for electrolytic dissociation in glass. We have also established the presence of a nonuniform distribution of the ionic components in the borate glasses in question. The compositional nonuniformity of these glasses can be considered to be experimentally established.

References

1.   R. L. Myuller, Zh. Fiz. Khim., 6:624 (1935).
2.   S. A. Shchukarev and R. L. Myuller, Zh. Fiz. Khim., 1:625 (1930).
3.   Ya. I. Frenkel (Frenkel'), Z. Phys., 36:652 (1926); W. Jost, J. Chem. Phys., 1:466 (1933); W. Schottky, Z. Phys. Chem., 29:355 (1935).
4.   R. L. Myuller, Phys. Z. Sowjetunion, 1:407 (1932).
5.   E. Seddon, E. J. Tippett, and W. E. S. Turner, J. Soc. Glass Technol., 16:459 (1932).

Chemical Features of Polymeric Vitreous Materials and the Nature of the Vitreous State

1. The vitreous state is a particular case of the amorphous state. It is realized under specific conditions on the transition from the liquid to the solid state. A high viscosity which leads according to Tammann [1] to the retardation of the crystallization process, in addition to forced cooling of the liquid through the melting temperature, facilitate glass formation. The high viscosity is determined physically, according to Kobeko [2], by the long relaxation time of the atomic processes in the liquid system. The long relaxation time makes it possible to obtain the thermodynamically nonequilibrium vitreous state which is stable at low temperatures. The reason for the different values of the relaxation time in substances differing one from another in chemical composition has not been explained theoretically. At the same time, the tendency to glass formation is intimately connected with the chemical nature of the material. In fact, we know that almost any normally vitreous chemical system can, according to A. A. Lebedev [3], be found under different conditions in the crystalline state, and that systems with typical elementary ionic structural units (halides, alkali and alkali earth oxides, etc.) cannot be obtained in the vitreous state.

The essential role in glass formation belongs to elements of the main subgroups of Groups III - VI. Zachariasen [4] noted the chemical differences between substances which crystallize readily and those which form glass with relatively greater ease. He proposed a classification of glass forming oxides based on Goldschmidt's crystal chemical data and put forward the well known network structure theory. Zachariasen's concept assumed a purely ionic structure for glass. The distinction he drew between the network-forming "ions" of the elements of the main subgroups of Groups III - V, and the network-modifying ions of the elements of the main subgroups of Groups I and II, was not based in principle on theoretical considerations. The nature of the chemical bonds and their role in retarding crystallization were not considered.

The role of localized covalent chemical bonds in the process of glass formation was emphasized by me at the First Conference on the Vitreous State. It was emphasized that "a knowledge of the fundamental structural unit and the nature of its chemical bonds is a key to the problem of the vitreous state... The tendency of a particular substance to form glass is associated with the prevalence of directed bonds with a small radius of action. These bonds are, primarily, powerful covalent bonds acting between the atoms of the high melting oxides of the Group III - V elements. In this case the number of bonds corresponds to the normal valency of these elements, and the resultant trigonal and tetrahedral orientation of the chemical bonds leads to the existence in oxygen compounds of typical crystalline lattices and amorphous atomic networks ... The covalent bonds in the atomic network (for example quartz) produce, at fairly low temperatures,

---

*R. L. Myuller, in: The Structure of Glass, Proceedings of the Third All-Union Conference on the Glassy State (1959), Izd. Akad. Nauk SSSR (1960), p. 61 [English translation: The Structure of Glass, Vol. 2, Consultants Bureau, New York (1960), p. 50].

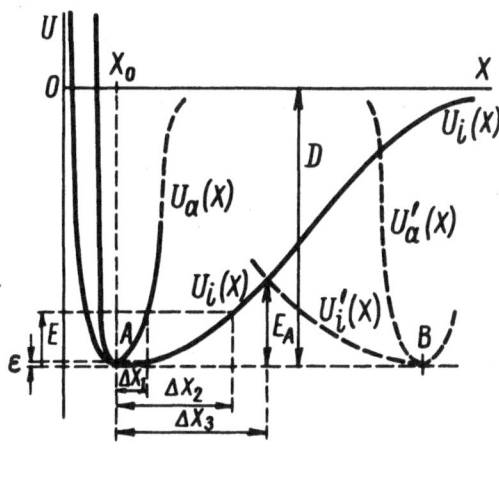

Fig. 1

atomic vibrations with an amplitude smaller than the amplitudes of the ionic vibrations in an ionic lattice (for example in sodium chloride). In this difference, I believe, lies the reason for the high viscosity and increased activation energy of atomic regrouping in substances with a tendency to form glass" [5].

In recent years a number of authors, notably Smekal, Dietzel, Syrkin, Tarasov, Stanworth, Sun, and Stevels, have also published work [6] which points to the essential role of the covalent components of the chemical bonds in glass formation, and Weyl has emphasized the importance of polarization of bonds. These authors paid particular attention to the high bond strength which inhibits their rupture and reorientation. Since crystalline non-glass forming substances frequently possess the same bond energies as glass forming substances, the physical nature of the peculiar position occupied by vitreous substances is still unexplained in principle. The subsequent treatment of this problem in the light of new data now allows us to investigate more deeply the nature of the vitreous state.

2. Atoms of the main subgroups of Groups I and II with valence s-electrons in the external shell are transformed in chemical compounds into ions with coulombic electrostatic fields evenly distributed throughout, which leads to the increased coordination number. Nonlocalized spherically and uniformly distributed chemical bonds are present between the oppositely charged ions. As the distance X between the ions increases the potential energy $U_i(X)$ (Fig. 1) increases in a flat curve ($U = -A/X$) and as a result the ionic chemical bond is long range and thus the deciding factor in the long range order in a crystalline lattice.

The wave mechanical behavior of the outer electrons in the atoms of compounds of the elements of the III - VI main subgroups is determined by the demand of integrality and for the saturation of the outer shell by eight electrons. In this case the filling of the outer shell by s^2p^6-electrons is realized by the creation of localized bonds from electron pairs. In this case the number of electrons in an atom determines the limit to the number of nearest neighbors. The strong binding of the electrons in localized bonds produces neither the ionic nor a metallic but a semiconducting character for such compounds. The localization of electron-pair bonds produce spatial limitation to the action of the chemical forces of the bond. Such spatially directed electron pair bonds, as distinct from ionic bonds, act at short range and therefore determine only the short range order. The number and direction of such bonds determine the coordination numbers and also the structural features of the crystal [7]. The electron-pair character of the bond is responsible for a steep rise in the potential energy $U_a(X)$ with an increase in the distance X between the atoms (Fig. 1). In this respect the covalent force constant k, the measure of the bond resistance, with an extremely small atom displacement becomes considerably bigger than the ionic force constant. To displace the covalently bonded atom a distance ΔX_1, an energy E is required, considerably higher than the value ε which is sufficient to displace an ion over the same distance ΔX_1, assuming equal values of the bond energy D in both cases. With the same energy E, the covalently bonded atom can be shifted only the small distance ΔX_1 while the electrostatically bonded ion shifts by $\Delta X_2 \gg \Delta X_1$. In any type of physical-chemical processes there are fairly large displacements of atoms and ions. Such displacements are made easier by the overlap of the potential field curves. In the case of ion displacement from one potential well A to a neighboring well B, an expenditure of the energy necessary for the complete rupture of the bond is not necessary. Only sufficient energy is required to surmount the potential barrier E_A at a critical distance ΔX_3. The latter is determined by the intersection of the curves $U_i(X)$ and $U_i'(X)$. This case of a long range ionic field relates to the translational motion of ions in an ionic dielectric.

In the case of a medium with short range covalent fields the conditions are different. In order to displace an atom the same distance ΔX_3 the full energy of covalent bond rupture is required as a result of the absence of the overlap of the potential field $U_a(X)$ and $U_a'(X)$ at large distances. In this case the lowering of the potential barrier is possible only when there is a direct overlap of the electron clouds of the valence bonds when the atoms are closely linked.

It follows, that when we examine the processes occurring in condensed systems it is necessary to take note of the values of the force constants characterizing the rigidity of the bonds between atoms and ions. This will allow for the extent of the possible spatial displacement of atoms and ions without complete rupture of the bonds. The widespread idea that the force constants of atoms and ions should not be of any interest to chemists [8] is mistaken. In fact the labile nature of the electrostatic bonds produces the increased fluidity of molten ionic salts, determines the marked ion diffusion in crystals and glasses, facilitates the rapid flow of ionic reactions, and produces an increased coefficient of expansion. All these processes in substances with an ionic structure occur with a relatively low activation energy, lowered because of the overlap of the long-range coulombic electrostatic fields created by the assembly of ions in the condensed system.

It is also reasonable to explain the lowered fluidity, the relatively low atomic diffusion coefficients and the low rates of chemical reaction, all of which are observed in polymeric vitreous systems, by the predominance of short range localized valence bonds between atoms. Increases in the activation energy for displacement and regrouping of atoms are associated with the inadequate overlap of the potential curves. This inhibits the transition of the structural elements in the melt to the state of complete readiness necessary for crystallization [9]. These kinetic features determine the "freezing" of the melt in the form of a glass.

The more extended the skeleton of the covalent atomic chains and atomic networks, the greater must be the difficulty in stacking them into geometrically ordered structures. We must note here that one thing is essential. Chain structures, like $[-S-]_n$, $[-Se-]_n$, $[-CH_2-]_n$ and so on, may produce slowly ordering interlacings (in the melt); in this case the decelerating role of the covalent bonds in crystallization is clear. The structures of diamond, silicon, germanium, indium arsenide, etc. although they are covalently bonded systems, even so cannot produce interlacings and do not form glasses. This is explained by the unequivocally tetrahedral structure which makes the disordered interlacings of covalently bounded chains impossible. At the same time boron, silicon, arsenic in their compounds with the main subgroup VI elements acquire bridges of the Si—O—Si type. The latter are chains which are interlaced in the molten state. This makes clear the fundamental importance of the role in glass formation of the group VI elements: O, S, Se, Te with s^2p^4 external electrons (Winter-Klein [10]). This group is of a unique type. The atoms of its elements are bonded covalently only with two neighboring atoms thus penetrating the network in interlacing chains (for example, by the penetration of the chains of one polyhedron though the internal opening of another polyhedron, Fig. 2a.)

The retardation of crystallization in substances which tend to form glass is thus explained by the slowing down of the atomic regrouping processes necessary to prepare the melt for the separation of crystals of a specified composition and structure. The decreased rate of such processes is determined by the high activation energy of displacement of covalently bonded atoms in the field of localized short range chemical forces which excludes the overlap (normal for ions) of fields at considerable distances. This is responsible for the long relaxation times of processes and for the considerable viscosity in such substances.

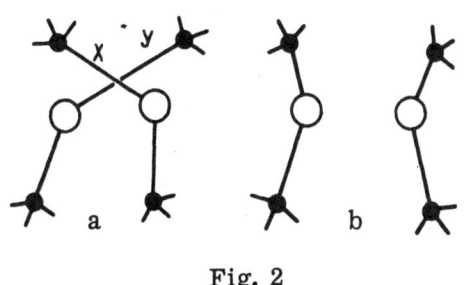

Fig. 2

As an illustration, we shall compare some data for easily crystallizable ionic compounds with the data for covalent compounds which readily form glasses. It is clear from Table 1 that, despite the low coordination number but thanks to the decreased interatomic distance, the packing density of the covalently bonded atoms is not essentially different from that of ions. Thus, relative to the higher valency, the free energy of the covalent lattice in quartz and boric oxide is higher than the energy of the ionic lattices of the alkali halides. At the same time, despite the close values of the bond energies (about 100 kcal/mole), the force constants of the covalent bonds are significantly higher than the force constants of ionic bonds.

Henceforth, when the covalent chemical bond is under consideration an allowance is made for the contribution of the ionic component to such a bond. The covalent contribution to the $B-O$ and $Si-O$ bond (40-60%), determined following Pauling from the electronegativity of the elements, is sufficient for them to exhibit the property of localized bond, namely that short range action which limits the atomic movement necessary for the flow of processes.

3. In order to confirm that we have correctly understood the nature of the vitreous state it is necessary to make systematic studies of the chemical kinetics of the processes which occur in vitreous media. We do not have such data available at present. However, we do have at our disposal sufficient experimental material on the temperature dependence of viscosity. These results can be used for preliminary confirmation of the valence concepts presented above of processes in molten glass. It is sometimes suggested, unreasonably, that the rate of structural reorganization is determined by the rate of breaking of chemical bonds [14]. In fact, to break bonds with an energy of 100 kcal/mole is so unlikely that we would be unable to explain the recorded rates of processes proceeding in melts. Assuming such a mechanism, at 1000°K there can be only about 10^{-20}% of the total number of bonds in borosilicates completely broken at any moment. Indeed, in glasses at 1000°K the activation energy for the viscosity is 24-37 kcal/mole, which corresponds to an instantaneous participation in the reorganization of about $10^{-5}-10^{-3}$% of all the chemical bonds [15]. It was noted above that, in the case of short range covalent bonds, an overlap of potential fields which reduces the activation energy is possible when there is direct overlap of the electron clouds due to the localized bonds being close together. In vitreous systems this is facilitated by the presence of bridges of divalent Group IV atoms providing the whole network structure with jointed, mobile bonds. Let us look at the physical model of the $SiO_{4/2}$ structural units (Fig. 3). The silicon and oxygen atoms are connected by the rigid bonds

TABLE 1. The Covalent-Ionic Characteristic of Solid NaCl, KCl, SiO₂, and B₂O₃

Chemical structural unit	Concentration, mole/cm³		Lattice energy kcal/mole [12] [13]	Chemical bond A - B	Bond energy, kcal/mole [8]	Interatomic distance, Å [12] d_{A-B}	$K \cdot 10^5$, dyn·cm⁻¹ [8]	C_p^T (6.4n)	C_p^e, 300°K	$P_e = \dfrac{C_p^T - C_e}{2.13}$	P_0
	of structural units [11]	of atoms or ions						cal/mole·deg [13]			
Na⁺Cl⁻	0.0370	0.0740	185	Na⁺—Cl⁻	98	2.8	1.2	12.8	11.9	0.4	0
K⁺Cl⁻	0.0267	0.0534	166	K⁺—Cl⁻	101	3.1	0.9	12.8	12.3	0.3	0
SiO₄/₂ {	0.0372	0.112	380	Si—O	104	1.54—1.58	—	19.2	10.6—10.8	3.9—4.0	4
				SiO	≈165	—	9.2				
—BO₃/₂ {	0.0532	0.133	311	B—O	110	1.36	—	16.0	7.3—7.4	4.0—4.1	4
				BO	≈185	—	13.6				

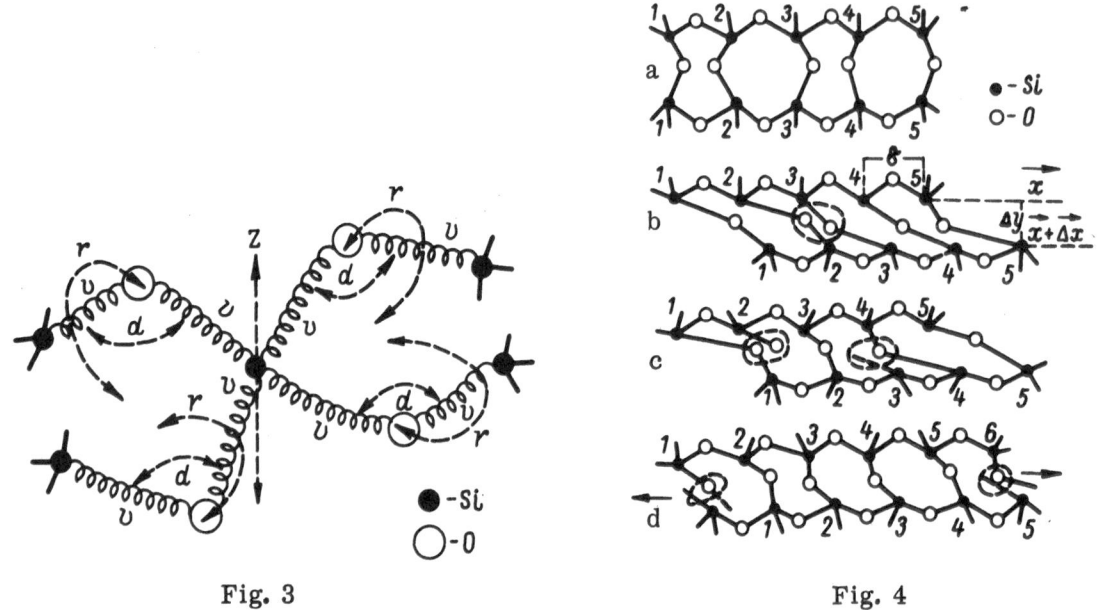

Fig. 3   Fig. 4

which have high force constants. Localized bonds interact relatively weakly at a distance and the elements of the structural unit have some degree of relatively free movement of a jointed character. The following modes of movement occur: deformational modes with a change in the angles d between the localized bonds; rotational-oscillatory motion of oxygen atoms along the arc r round the Si ... Si axis normal to the plane of Fig. 3; and the oscillatory motion z of the central silicon atom.

The combination of these movements allows the overlap of the localized electron clouds of the bonds (see Fig. 2(a)) on the application of potential fields and the consequent lowering of the activation energy. The overlap of covalent bonds can thus exist, the transition from scheme (a) to scheme (b) occurs (Fig. 2), and in the same way the structural reorganization is realized without the prior complete rupture of the bonds. Thus, the diffusion of atoms, the migration of bonds, and the flow of the substance can occur at a measurable rate.

In measuring the viscous flow of a liquid it has been observed that the establishment of a steady rate of the relative displacement of layers of atoms was preceded by an induction period with a low activation energy extension of the network, with a change in the angles between localized bonds: the transition from scheme (a) to scheme (b) occurred (Fig. 4). After the extension of the network has been achieved the interaction between rotationally vibrating oxygen bridges begins (Fig. 4(b)). If the activation energy for such an interaction between, for example, $Si(2)-O-Si(2)$ and $Si(3)-O-Si(3)$ is available, it results in the switching of bonds with the development of a new bridge $Si(3)-O-Si(2)$ and the radicals $Si(2)-O-$ and $-Si(3)$. The latter interact easily with the neighboring bridges $Si(1)-O-Si(1)$ and $Si(4)-O-Si(4)$ (Fig. 4(c)), and thus produce a switch over to different sides (Fig. 4(d)). The mean direction of switching and of the consequent relative displacements of silicon atoms, is determined by the small decrease in the activation energy under the influence of the external directional force, and this produces the gradient in the layers, $\Delta x/\Delta y$ (Fig. 4(b)) [15].

The viscosity equation [15] was theoretically derived from this valence model for the viscous flow of glass. The theoretical value of the pre-exponential term was found to agree satisfactorily with the experimentally determined values. A critical analysis of the experimental data for the temperature dependence of the viscosity of glasses shows that, at about 150-250°C above the melting point, the free energy of activation of the viscosity is temperature independent.

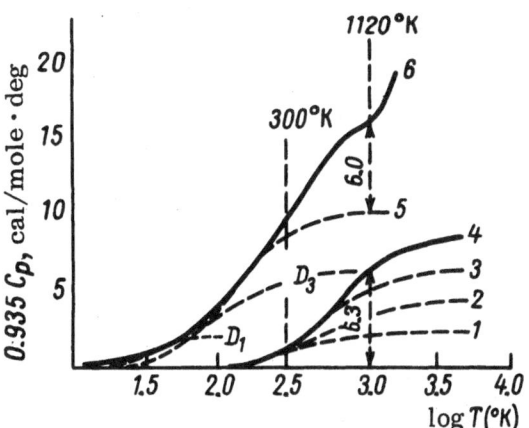

Fig. 5. Einstein functions: 1) E_1 (1190/ T), 830 cm^{-1}(12 μ); 2) E_2 (1490/ T), 1040 cm^{-1} (9.6 μ); 3) E_3 (1670/ T), 1170 cm^{-1} (8.5 μ); 4) E_4(1880/ T), 1290 cm^{-1} (7.8 μ); 5) Debye functions t_5(568/ T), 400 cm^{-1}(25 μ); 6) Experimental curve of the thermal capacity of SiO$_{4/2}$ at constant volume.

Such an anentropic activation energy is not observed at temperatures below the melting point [16]. Without going into the analysis of this complex phenomenon we may note that it is associated with the features of a covalent movement of atoms in covalently bonded jointed structures.

4. Thermal capacity is a measure of the increase with temperature of the intensity of the atomic and ionic vibrations in solids. The singly charged Na$^+$, K$^+$, and Cl$^-$ ions connected by labile, low force constant coulomb forces are characterized by the relatively small size of the vibrational quanta. Thus, in the ionic crystals NaCl and KCl, the vibrational wave numbers lie below 400 cm^{-1} [12]. In covalently bonded B$_2$O$_3$ and SiO$_2$ the wave numbers reach 1440 and 4650 cm^{-1} respectively. Therefore, in covalently bonded structures the atomic thermal capacities are lower than those in ionic structures (see Table 1) [13]. Taking the average value for the atomic thermal capacity at constant pressure with the energy uniformly distributed among all the vibrational degrees of freedom as being 6.4 cal/ g-atom · deg [17], we shall calculate the thermal capacity for solids with a uniform dis-

tribution from C_p^t = 6.4 n cal/mole·deg, where n is the number of atoms and ions comprising the chemical structural unit. In ionic NaCl and KCl at T $\approx$ 300°K the experimental thermal capacity C_p^e corresponds to the theoretical thermal capacity C_p^t - corresponding to the uniform distribution of energy among all the oscillatory degrees of freedom of the ions. In B$_2$O$_3$ and SiO$_2$, both in the crystalline and in the vitreous state, there is a significant decrease in the thermal capacity by the amount $C_p^t - C_p^e$. This decrease corresponds roughly to the exclusion from the energy partition of four vibrational degrees of freedom $(C_p^t - C_p^e)/2.13$ for each structural unit. This corresponds to the four valence bonds which exist in the SiO$_{4/2}$ structural units and =O: → BO$_{3/2}$ (see the coordination of oxygen atoms in [18]).

In solid lattices and networks of the B$_2$O$_3$ and SiO$_2$ type with jointed bonding of atoms by localized bonds with high force constants it is conceivable that the modes of free vibrational motion are divided into two groups.

Group I comprise the acoustic and low-frequency optical vibrations consisting of atomic displacements with small force constants analogous to those observed in NaCl and KCl. Assuming a sufficiently weak interaction between the localized bonds in quartz-like lattices and networks we can assume that the z-, d-, and r- oscillatory displacements of silicon and oxygen atoms are to be included in Group I (only five degrees of freedom of motion in the structural unit).

Group II include the higher frequency optical vibrations. The valence oscillatory displacements with larger force constants of the atoms in Si—O will be included herein. There are four such degrees of freedom in the structural unit.

The inclusion, at higher temperatures, of the latter high frequency valence vibrations is statistically possible when there is jointed bonding of the atoms. This must facilitate damping at low temperatures of the large quanta valence vibrations by the whole jointed bonded system

TABLE 2. Jointed Network Structures (Y ≥ 3.5; ξ ≥ 0.5)

Chemical structural units (ξ)	Modification	$C_p^T = 6.4 n$	C_p^e [20] (298—300°K)	$P_e = \dfrac{C_p^T - C_p^e}{2.13}$	P_0	P_e/P_0
		cal/mole · deg				
SiO₂	Quartz	19.2	10.6—10.8	3.9—4.0	4	0.97—1.00
	Cristobalite	19.2	10.7	4.0	4	1.00
	Glass	19.2	10.6	4.0	4	1.00
BO₁.₅	"	16.0	7.4	4.0	4	1.00
	Crystalline	16.0	7.4	4.0	4	1.00
AlO₁.₅	Alumina	16.0	9.4	3.1	3	1.03
Al₂SiO₅	Andalusite	51.2	32.1	9.0	10	0.90
	Sillimanite	51.2	30.4	9.8	10	0.98
	Disthene	51.2	29.6—30.4	9.8—10.1	10	0.98—1.01
	Kyanite	51.2	28.0	10.9	10	1.09
LiAlSi₄O₁₀ (0.20)	Petalite.	102.4	62	18.9	20	0.94
Li₂Al₂Si₄O₁₂ (0.33)	Spodumene	128.0	76	24.4	24	1.02
NaAlSi₃O₈ (0.25)	Albite.	83.2	47.3—48.7	16.2—16.9	16	1.01—1.05
	Glass	83.2	49.4	15.8	16	0.99
KAlSi₃O₈ (0.25)	Orthoclase.	83.2	45.9—48.2	16.4—17.5	16	1.02—1.09
	Microcline.	83.2	49.5	15.8	16	0.99
	Glass	83.2	46.3—50.9	15.1—17.2	16	0.95—1.07
BeAl₂Si₂O₈ (0.50)	Beryl	83.2	51	15.1	16	0.95
CaAl₂Si₂O₈ (0.50)	Anorthite	83.2	49.5 50.5	15.3—15.8	16	0.96—0.99
	Glass	83.2	49.4—49.7	15.7—15.9	16	0.98—0.99
Fe₃Al₂Si₃O₁₂ (1.20)	Garnet	128.0	87	19.2	20	0.96
3(NaAlSiO₄·2SiO₂)2CaAl₂Si₂O₈ (0.35)	Crystalline.	416.0	242—246	80—82	80	1.00—1.03
Na₂B₄O₇ (0.50)	"	83.2	49.3	15.9	16	1.00
K₂B₄O₇ (0.50)	"	83.2	51.3	15.0	16	0.94

with its labile small quanta vibrational components. Thus, despite the absence of a break in the experimental curve of the temperature dependence of the thermal capacity of quartz in the region of 300°K, on the basis of the infra red absorption spectrum [12,19] we can separate the vibrations with wave numbers ≥ 800 and ≤ 400 cm⁻¹. The curves for partial thermal capacity (see Fig. 5) are in satisfactory agreement, within 5%, with the experimental values.

The values of the thermal capacity in covalent bonded structures of various chemical compositions and structures (see Table 2) deserve some attention. Without exception when Y ≥ 3.5 according to Stevels (or when there is ionization ξ of not more than 50% of the structural units), the thermal capacity at about 300°K has a value that would correspond to the uniform partition of energy among all the degrees of freedom left after allowing for the number of valence degrees, P_0. The ratio of the experimental value of the effective number of the excluded degrees of freedom P_e to the number of valence bonds in the structural chemical unit is close to unity.

The regular decrease in the thermal capacity of vitreous substances in relation to the number of covalent bonds requires serious investigation. It requires the development of a theory of thermal capacity of solids with atomic structures with jointed bonding by localized covalent bonds which precludes the use of linear and homogeneous equations.

5. Organic polymeric systems of the (−CH₂−) type are typical examples of covalently bonded chain structures.

In such structures physical chemical reorganizations also occur without complete bond breaking. The energy of the covalent bonds C−C is 60-80 kcal/mole [21] and such bonds break at appreciable rates only at 750-1000°K. Therefore the structural reorganizations observed in organic polymers must also be realized by the switching of valence bonds. Like the inorganic networks and chain structures (Fig. 6, I) the organic polymeric units (Fig. 6, II) on changing from one structure (Fig. 6(a)) to others (Fig. 6, (c), (d)) pass through an activated complex state A, which is characterized by the mutual overlap of valence electron clouds of neighboring

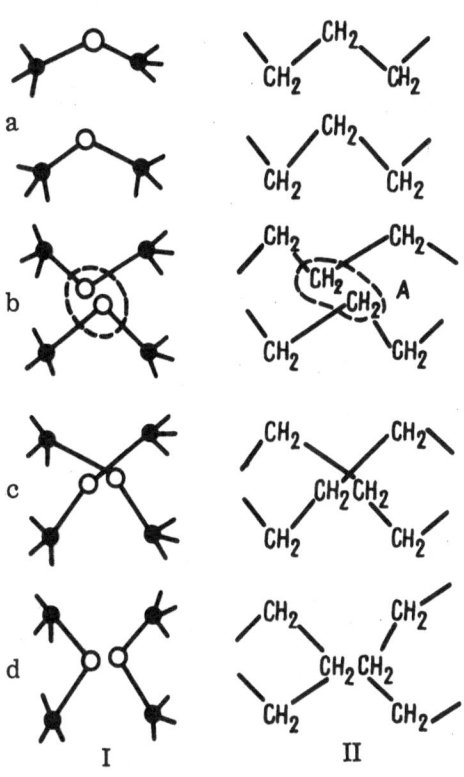

I   II

Fig. 6

units (Fig. 6, (b)) which enable the bonds to regroup without a preliminary complete rupture.

In the region 20–50°C we observe an abrupt jump in the thermal capacity of such systems and at low temperatures we note the transition to a stable vitreous state with the exclusion of torsional–vibrational degrees of freedom from the energy distribution [22].

In addition to the covalently bonded polyatomic chemical substances there are molecular systems producing significantly less stable glasses. The latter are polymolecular structures bonded by weak polar van der Waals forces. Among the compounds which produce this type of glass are ethyl and propyl alcohol, glycol, propyleneglycol, and glycerine [23]. In these compounds in the liquid state the jump in the thermal capacity is observed in the region of 100–200°K. The vitreous state in such molecular liquids owes its existence to the polar short range forces of interaction ($U_p = -B/X^3$) which produce increased activation energies for the breaking of the polar bonds between the molecules at these low temperatures [22].

6. The analysis of vitreous materials, both of high melting aluminoborosilicates and of the sulphides and selenides of Group V elements and of the lower melting organic polymeric systems and associated polar liquids, leads to the following conclusion:

The phenomenon of glass formation is the result of the slowing down of the crystallization process owing to the increased activation energy of the structural reorganization necessary as a preparation for crystallization, which involves the redistribution of particles and their mutual orientation. The activation energy in vitreous substances is high as a result of the bonding of atoms, of highly electrovalent ions (the "Invert Glasses" of Stevels [25]) or molecules, by the short range forces (covalent, polarization, or dipolar respectively) of the chemical bonds.

The real problem in studying the vitreous state is to set up and develop, in the very near future, experimental studies of the kinetics of the physical chemical processes in glasses, liquids, viscous and solid vitreous media and to develop a statistical theory of the corresponding atomic–molecular processes.

References

1. G. Tammann, Kristallisieren und Schmelzen, Leipzig (1903).
2. P. P. Kobeko, Amorphous Substances, Izd. Akad. Nauk SSSR, Moscow (1952).
3. A. A. Lebedev, Tr. Gos. Optich. Inst., Vol. 2, No. 10 (1921); Zh. Russk. Fiz.-Khim. Obshchestva, 50:1, 57 (1921); Izv. Akad. Nauk SSSR, ser. fiz., 4:584 (1940).
4. W. H. Zachariasen, J. Amer. Chem. Soc., 54:3841 (1932); Phys. Rev., 47:277 (1935).
5. R. L. Myuller, Izv. Akad. Nauk SSSR, ser. fiz., 4:607 (1940) ● this volume, p. 170; Zh. Prikl. Khim., 13:479 (194); The Vitreous State and the Electrochemistry of Glass (Doctoral Dissertation), Leningrad (1940).
6. A. Smekal, Angew. Chem., 55:235 (1942; Über die Existenzbedingungen von Glaszustanden. Zur Struktur und Materie der Festkörper, Berlin (1952), p. 223; A. Dietzel, Z. Elektrochem.,

48:9 (1942); Naturwiss., 31:25 (1943); Glastechn. Ber., 22:41 (1948); Ya. K. Syrkin and M. E. Dyatkina, The Chemical Bond and the Structure of Molecules, Moscow (1946), p. 400; V. V. Tarasov, Dokl. Akad. Nauk SSSR, 46:117 (1945); 58:577 (1947); Zh. Fiz. Khim., 24: 111 (1950); J. E. Stanworth, J. Soc. Glass Technol., 30:56 (1946); Glastechn. Ber., 23:299 (1950); K. H. Sun, J. Amer. Cer. Soc., 30:277 (1947); M. L. Huggins and K. H. Sun, J. Soc. Glass Technol., 28:463 (1944); J. Stevels, Verr. et Refract., 7:91 (1953); T. Forland and W. A. Weyl, J. Am. Cer. Soc., 32, 269 (1949); 33:186 (1950); W. A. Weyl, Colored Glasses, Sheffield (1951).

7. F. Hund, Z. Electrochem., 61:891 (1957).
8. T. Cottrell, The Strength of the Chemical Bond, London (1954).
9. N. V. Belov, in: The Structure of Glass, Vol. 1, Consultants Bureau, New York (1958).
10. A. Winter-Klein, Verr. et Refract., 7:147 (1955); IVth Congress International du Verre, VIII, 1 (1956).
11. Chemist's Handbook, Vol. II, Goskhimizdat, Moscow—Leningrad (1951).
12. Landolt-Börnstein, Zahlenwerte und Funktionen aus der Physik, Chemie, Astronomie, Geophysik, und Technik, Vol. 1, Atom- und Molekular-physik, Vol. 4, Kristalle, Berlin—Göttingen —Heidelberg (1955).
13. F. D. Rossini, D. D. Wagman, W. H. Evans, S. Levin, and I. Jaffe, Selected Values of Chemical Thermodynamic Properties, Nat. Bur. Stand. Circ., No. 500 (1952).
14. O. L. Anderson and D. A. Stuart, Ind. Eng. Chem., 46:154 (1954).
15. R. L. Myuller, Zh. Prikl. Khim., 28:363, 1077 (1955).
16. K. Arndt, Z. Electrochem., 13:500 (1907); K. S. Evstrop'ev, M. M. Skornyakov, and B. A. Pospelov, in: The Physical Chemical Properties of the Triple Systems: NaO—PbO—SiO, Izd. Akad. Nauk SSSR, Moscow—Leningrad (1949); A. A. Leont'eva, Zh. Fiz. Khim., 24: 798 (1950); V. T. Slavyanskii, in : The Vitreous State, Izd. Akad. Nauk SSSR, Moscow—Leningrad (1960), p. 328.
17. R. Fowler and E. Guggenheim, Statistical Thermodynamics [in Russian], I. L. Moscow—Leningrad (1949).
18. S. Krogh-Moe, Glastechn. Ber. 32K:VI-18 (1959).
19. R. B. Barnes, Phys. Rev., 39:562 (1932); W. Stein, Ann. Phys., 36:462 (1939); K. Kol'raush, Combination Scatter Spectra, Russian ed., IL. Moscow (1952), p. 367.
20. W. Eitel, Thermodynamical Methods in Silicate Investigation, New Jersey (1952).
21. N. N. Semenov, Some Problems in Chemical Kinetics and Reaction Capabilities, 2nd ed., Moscow (1958).
22. R. L. Myuller, Zh. Fiz. Khim., 28:2189 (1954).
23. W. Eitel, Physical Chemistry of Silicates, Chicago (1954), pp. 250, 276; G. S. Parks, S. B. Thomas, and W. A. Gilkey, J. Phys. Chem., 34:2028 (1930).
24. S. W. Hawley, J. Amer. Chem. Soc., 29:1011 (1907); R. Frerichs, Fiz. Rev., 72:12 (1947); J. Opt. Soc. Amer., 43:1153 (1953); W. A. Fraser, J. Opt. Soc. Amer., 43:823 (1953); G. Dewulf, Rev. Optique, 33:513 (1954); N. A. Goryunova and B. T. Kolumietz, Zh. Tekhn. Fiz., 25:2070 (1955).
25. W. A. Weyl and E. C. Marboe, Glastechn. Ber., 32K:VI-1 (1959); H. J. L. Trap and J. M. Stevels, Glastechn. Ber., 32K:VI-32 (1959).

Thermal Ionization and
the Current Carrier Mobility in Glass*

Characteristically, vitreous substances have a three-dimensional skeleton of covalently bonded atoms as in silica ($SiO_{4/2}$) [1]. Such a skeleton, while stabilizing the material mechanically and chemically, also separates and thus weakens the interaction of the polar groups. These groups consist of ionized oxygen, boron or aluminium atoms connected directly by coulomb forces to the metallic cations ($M^+O^-SiO_{3/2}$, $M^+B^-O_{4/2}$, and $M^+Al^-O_{4/2}$). In contrast to crystals with typical ionic structures, the short-range forces of the covalent bonds predominate in the vitreous bodies in question and determine the short-range order of the atoms in such structural units as $SiO_{4/2}$, $BO_{3/2}$, $B^-O_{4/2}$, etc.

The absence of clearly expressed long range coulomb interaction forces determines the weak dependence of the energy of these materials on the degree of geometric order in their structure. Table 1 illustrates this point. The values of the energy of atomization are shown for several substances in both the crystalline and in the vitreous states. These data were obtained from the standard values for the enthalpy at 25°C and the heat of sublimation of solids [2, 3, 4] according to the equation

$$\Delta H^0_{298}(xA,\ yB,\ zC\ \ldots) = -\Delta H^0_{298}(A_xB_yC_z\ \ldots)\text{solid} +$$
$$+ x\Delta H^0_{298}(A)\text{gas} + y\Delta H^0_{298}(B)\text{gas} + z\Delta H^0_{298}(C)\text{gas} + \ldots$$

TABLE 1. Atomization Energy of Solids, in the Vitreous and in Various Crystalline States

Chemical composition	Atomization energy, kcal/mole		Difference %
	Crystals	Glass	
SiO_2	411.8 (quartz) 411.4 (cristobalite) 411.2 (tridymite)	408.9	0.7
B_2O_3	674.0	669.6	0.7
P_4O_{10}	1612.8	1626.8	0.9
Na_2SiO_3	593.0	590.0	0.5
$K_2O \cdot Al_2O_3 \cdot 4SiO_2$	2662.0 (leucite)	2650.6	0.4
$K_2O \cdot Al_2O_3 \cdot 6SiO_2$	3484.0 (microcline) 3510.0 (adularia)	3448.0	1.0 1.8
$CaO \cdot 2B_2O_3$	1650.1	1637.5	0.8
$12CaO \cdot 7Al_2O_3$	4617.0	4585.0	0.7
$2CaO \cdot Al_2O_3$	704.0	695.0	1.3
$3CaO \cdot Al_2O_3$	861.0	848.0	1.5
Se	49.4	48.3	2.3

*R. L. Myuller, in: The Electrical Properties and the Structure of Glass, Izd. "Khimiya" (1964), p. 15 [English translation: The Structure of Glass, Vol. 4, Consultants Bureau, New York (1965), p.64]

TABLE 2. The Energy of the Chemical Bonds in Vitreous Solids
from Atomization Energy Data

Chemical composition	Structural unit	Atomization energy, kcal/mole	Chemical bond	Bond energy, kcal/mole
SiO_2	$SiO_{4/2}$	409	Si—O	102
B_2O_3	$BO_{3/2}$	335	B—O	112
P_4O_{10}	$O^-P^+O_{3/2}$	407	P—O	81
Na_2SiO_3	$\left(Na_2^+O_{2/2}^-\right)(SiO_{4/2})$	590	Na^+O^-	91
$K_2O \cdot Al_2O_3 \cdot 4SiO_2$	$(K^+Al^-O_{4/2})_2 (SiO_{4/2})_4$	2650	K^+Al^-	98
$K_2O \cdot Al_2O_3 \cdot 6SiO_2$	$(K^+Al^-O_{4/2})_2 (SiO_{4/2})_6$	3448	K^+Al^-	88
S	$SS_{2/2}$	—	S—S	55 [5]
Se	$SeSe_{2/2}$	94	Se—Se	47
As_2S_3	$AsS_{3/2}$	—	As—S	51 [5]
As_2Se_3	$AsSe_{3/2}$	—	As—Se	45 [6]
$GeSe_2$	$GeSe_{4/2}$	—	Ge—Se	49 [6]

A comparison of the values of the atomization energy of these solids in the crystalline and vitreous states indicates that there is little difference even when they have a significant content of polar structural units. Since the long range geometrical order does not provide any essential contribution to the atomization energy, the latter can, to a first approximation, be determined by the short range order atomic interactions in the $SiO_{4/2}$, $BO_{3/2}$, $AlO_{3/2}$, etc. structural units. By attributing the atomization energy basically to localized chemical bonds we can provisionally estimate the value of the energy required to break the respective chemical units [4]. Table 2 shows the values obtained in this way for the energy of the chemical bonds of the localized units. It was assumed that the values of the bond energies are close in the structural units $SiO_{4/2}$ and $Al^-O_{4/2}$ (approximately 102 kcal/mole of the bonds).

It is worth comparing such thermodynamically determined values of the energy of covalent, and in effect homopolar [11] bonds, with the values of their ionization energy determined from the optical absorption edge, and from the temperature dependence of the conductivity in chalcogenide glasses in the Ga—Ge—As—Se—S system under conditions in which there is continuous conductivity ($\log \beta \approx 0$) [19]. Such a comparison is made in Table 3 using the experimental data for the conductivity of vitreous Se [9], $AsS_{2.5}$ [5], $AsSe_{1.5}$ [7, 9], $GeSe_4$, and $AsGe_{1.5}Se_{4.5}$ [6,10]. It is clear that in these conditions of free continuous conductivity the ionization energy of the valence bonds in these glasses is very similar to the energy of these bonds and lies within the limits 1.7–2.2 eV. It also follows that the activation energy of the translational displacement of the free current carriers (electrons and vacancies) $\varepsilon_a = 0.5 (\varepsilon_\sigma - \varepsilon_\lambda)$ [9] does not exceed 0.2 eV and is approximately 10% of the ionization energy.

The elements B, Al, Si, P, and O which are the basis of inorganic oxide glasses (see Fig. 1) directly adjoin the elements of the main sub-Groups III–VI in the periodic table. It is characteristic of these glasses to have a high value for ionization energy of the valence bonds, in excess of 3.5 eV according to the data in Table 2. In such glasses the electron conductivity is exceeded by the ionic conductivity, as the energy of electrolytic dissociation of the polar groups M^+X^- is about 1–2 eV. Although the breaking energy of an ionic bond in such polar groups is around 4 eV (Table 2) and in fact differs little from the ionization energy of the covalent bonds, nevertheless the energy of electrolytic dissociation is low and is ~ 1–2 eV. This is the result

TABLE 3. The Ionization Energy of the Covalent Bonds and
the Activation Energy of Displacement of Free Current Carriers in Chalcogenide Semiconducting Glasses (in eV)

Chemical bond	Bond energy	Optical absorption edge $\varepsilon\lambda$	Conductivity energy ε_σ	Activation energy ε_α
Se—Se	2.0	1.76 [7, 8]	1.7 [8, 9]	0.05
As—S	2.2	2.2 [5]	2.2 [5]	0.05
As—Se	2.0	1.7 [7, 9]	1.7 [7, 9]	0.05
Ge—Se	2.1	1.8 [6, 10]	2.2 [6, 10]	0.2

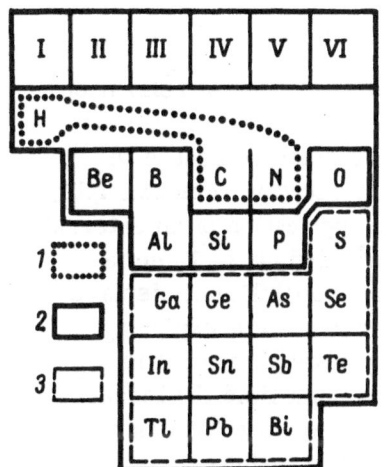

Fig. 1. Elements of the main
sub-groups in the periodic
table: 1) forming organic polymers; 2) inorganic vitreous
polymers; and 3) non-oxide
semiconductors.

of the high value of the quasisolvation energy, the energy of
interaction of the free cations obtained on dissociation and of
the negatively charged vacancies with the surrounding dipoles of the polar groups [12]. In this case dipole orientation
is achieved by the displacement of the cations inside the polar structural units [13]. The Frenkel'—Onsager calculation
indicates that in alkali borate glasses with a dielectric constant around 18 the energy necessary for the escape of a cation from the $Na^+B^-O_{4/2}$ polar structural unit is 85% recov-
ered from the energy of the quasisolvation effect [12]. In
this case the energy of electrolytic dissociation ε_i decreases to 2.1 eV while the activation energy ε_α of the translational displacement of a free cation is about 0.2 eV.

Thus, in both electron and ionic conduction in glasses
we observe values for the activation energy for the displacement of free current carriers which are low in comparison
with the ionization energy of a covalent bond or with the energy
of electrolytic dissociation of a polar unit. The value of the
energy ε_α (energy of conductivity) which is derived from the
temperature dependence of the conductivity of glasses is de-
termined, to a first approximation, by the ionization energy ε_i of the structural unit which fixes the concentration of current carriers. Thus, for example, when ε_i = 30 kcal/mole and ε_α =
5 kcal/mole, we have at 500°K a proportion of ionized structural elements $\alpha_i = (-\varepsilon_i/2kT)$
10^{-7}, while the ratio of current carriers in translational motion to the total number of "free"
carriers is $\alpha_\alpha = \exp(\varepsilon_\alpha/kT) \approx 10^{-2}$. In other words, $a_i \approx 10^{-5} \alpha_\alpha$.

The introduction of oxides of transition elements with d-electrons into low alkali oxide
glasses is known to be accompanied by an abrupt decrease in the energy of formation of free
electrons (ε_i < 0.5 eV). This produces in such glasses an easier overtaking of the decreased
ionic conductivity by the electron conductivity.

The principles noted here for oxygen and non-oxygen containing inorganic glasses must
also occur in organic polymers. The latter are directly juxtaposed in composition and properties to the inorganic polymers which are in essence vitreous substances (see Fig. 1).

In view of the above considerations it is rather unlikely that there was 100% electrolytic
dissociation of the Na^+O^- sodium polar groups in the molten silicate glass studied by A. F.

TABLE 4.

Structural element	Concentration of free current carriers		Increase per 100°
	at 1500°K	at 1600°K	
Na^+O^-	0.02 n	0.025 n	0.005 n
$Si-O$	$3.2 \cdot 10^{-6}$ n	$6.3 \cdot 10^{-6}$ n	$3 \cdot 10^{-6}$ n

Borisov and V. I. Zadumin. Starting from the values they obtained for the energy of the $Si-O$ (75 kcal/mole) and $Na^+ - O^-$ (23 kcal/mole) bonds, we should expect the concentration of free current carriers shown in Table 4 (n is the concentration of Na^+O^- $SiO_{3/2}$ structural elements).

It follows from Table 4 that at about 1300°C only 2-2.5% of the Na^+O^- polar structural units are dissociated, while in only 0.01% of the structural units are the $Si-O$ bonds broken.

It also follows that in alkali oxide glasses we have to distinguish two types of relaxation loss: those caused a) by the localized displacements within the limits of each of the dipole structural units of a large number of bound alkali cations and b) by the displacement of the small number of the free cations which determine the steady state conductivity. For example, when $\varepsilon_i \approx 25$ kcal/mole and $\varepsilon_a \approx 5$ kcal/mole, immediately after the application of an external field at 500°K the number of displacements of bonded cations in 1 cm³ of glass in 1 sec is $n\nu \exp(-\varepsilon_a/kT)$ sec^{-1}, i.e., for all the cations in the glass the mean probability of displacement is $f_a = \exp(-\varepsilon_a/kT) \approx 10^{-2}$. The number of displaced free cations is $n_i\nu \exp(-\varepsilon_a/kT) = n\nu \exp(-0.5\varepsilon_i + \varepsilon_a/kT)$, i.e., for all the cations the mean probability of displacement is $f_b = \exp(0.5\varepsilon_i + \varepsilon_a/kT) \approx 10^{-8}$. Immediately after the application of an external field the losses caused by the limited displacements of a significant number of cations may be 10^6 times greater than the losses produced by the displacement of a small number of free cations.

With a high alkali oxide concentration in the glass it is also possible to have cation displacement limited by the dimensions of amicron dispersed associated polar inclusions in the nonpolar medium [14].

The generalizations of L.A. Grechanik, E. A. Fainberg, and I. N. Zertsalova relating to the experimental data for the conductivity of a large number of glasses studied by various authors is interesting. The graph they obtained indicated the existence of a regular dependence of the log of conductivity log σ on the energy of conduction ε_σ. Particularly at 200°C this dependence can be expressed, to a first approximation, as

$$\log \sigma \approx 2.4 - 5.6\varepsilon_\sigma, \tag{1}$$

where σ is expressed in $\Omega^{-1} \cdot cm^{-1}$ and ε_σ in eV.

The statistical pre-exponential term in the expression for conductivity

$$\sigma = \sigma_0 \exp\left(-\frac{\varepsilon_\sigma}{2kT}\right) \tag{2}$$

can be calculated theoretically. Both in the case of ionic conducting and semiconducting glasses a constant is obtained for the conductivity modulus

$$\log \frac{\sigma_0}{n} \approx \log \frac{\sigma_0}{[v]} \approx 4, \tag{3}$$

where n and [v] are the concentrations in mole/cm^3 of the polar structural groups and the co-valent bonds, respectively. In the glasses studied by Grechanik, Fainberg, and Zertsalova the values of n and [v] are to an order of magnitude 10^{-2} mole/cm^3. If we take the value of 2 kT at 200°C (473°K) as $8.15 \cdot 10^{-2}$ eV, then after substituting the provisional value of log $\sigma_0 \approx 2$ derived from Eq. (3) in Eq. (2) we obtain

$$\log \sigma \approx 2 - \frac{\varepsilon_\sigma}{2.3 \cdot 8.15 \cdot 10^{-2}} = 2 - 5.3\varepsilon_\sigma. \tag{4}$$

As we see, the theoretical expression, Eq. (4), for the functional dependence of log $\sigma(\varepsilon)$ at 200°C and n $\approx$ [v] $\approx 10^{-2}$ agrees satisfactorily with the empirically established relationship of Eq. (1).

We must also note, in conclusion, that it has been recently established that the empirical dependence of conductivity energy ε_σ on the concentration n of the polar structural units in electron conducting glasses containing polar groups is given by

$$\varepsilon_\sigma \cdot n^{1/4} = \text{const}, \tag{5}$$

which had previously been established for ion-conducting silicate glasses [15].

References

1. R. L. Myuller, in: The Structure of Glass, Vol. 2, Consultants Bureau, New York (1960), p. 50, ● this volume, p. 178.
2. W. M. Latimer, The Oxidation States of the Elements and Their Potentials in Aqueous So-lution, 2nd ed., Prentice-Hall, New York (1952).
3. F. D. Rossini, D. D. Wagman, W. H. Evans, S. Levin, and I. Jaffe, Selected Values of Chem-ical Thermodynamic Properties, Nat. Bur. Stand. Circ., (1952), No. 500.
4. Vedeneev, L. V. Gurvich, V. N. Kondrat'ev, V. A. Medvedev, and E. L. Frankevich, The Energy of the Breaking of Chemical Bonds, the Potentials of Ionization and Electron Af-finity, Izd. Akad. Nauk SSSR, Moscow (1962).
5. R. L. Myuller, L. A. Baidakov, and Z. U. Borisova, Vestn. Leningr. Gos. Univ., No. 22, Iss. 4, p. 77 (1962).
6. R. L. Myuller, L. A. Baidakov, and Z. U. Borisova, Vestn. Leningr. Gos. Univ. No. 10, Iss. 2. p. 94 (1962), ● this volume, p. 143.
7. L. A. Baidakov, Z. U. Borisova, and R. L. Myuller, Zh. Prikl. Khim., 34:2446 (1961), ● this volume, p. 133.
8. T. S. Moss, Photoconductivity in the Elements, London (1952).
9. R. L. Myuller, Zh. Prikl. Khim., 35:541 (1962), ● this volume, p. 121.
10. Z. U. Borisova, R. L. Myuller, and Chin. Ch'eng-Ts'ai, Zh. Prikl. Khim., 35:774 (1962).
11. E. Mooser and W. B. Pearson, Nature, 190:406 (1961).
12. R. L. Myuller, Zh. Tekhn. Fiz., 25:1567 (1955), ● this volume, p. 43.
13. R. L. Myuller, Zh. Tekhn. Fiz., 25:1566 (1955).
14. R. L. Myuller, Zh. Tekhn. Fiz., 26:2614 (1956), ● this volume, p. 79.
15. A. V. Danilov and R. L. Myuller, Zh. Prikl. Khim., 35:2012 (1962).

List of R. L. Myuller's Scientific Papers

In the following list the papers have been grouped in sections corresponding to the main independent directions of R. L. Myuller's scientific work. In each section, the papers have, as far as possible, been arranged chronologically. Articles which are closely related textually are placed under the same number but with a distinguishing index (Editor's note).

Ionic Conduction in Glasses

1.   S. A. Shchukarev and R. L. Myuller, "A study of the electrical conductivity of glasses in the system $B_2O_3-Na_2O$," Zh. Fiz. Khim., 1:625 (1930).

1a. S. A. Shchukarev (Schtschukarew) and R. L. Myuller (Müller), "Untersuchung der elektrischen Leitfähigkeit von Gläsern," Z. Phys. Chem., A150(5/6):439 (1930).

1b. S. A. Shchukarev and R. L. Myuller, "The specific conductivity of borate glasses" (Abstract, Report of the Russian Phys. Chem. Soc.), Zh. Russk. Fiz.-Khim. Obshchestva Chast' Khim., 62(1):230 (1930).

2.   R. L. Myuller (Müller), Das Wesen der Ionenleitfähigkeit von Gläsern," Phys. Z. Sowjetunion, 1(3):407 (1932).

3.   R. L. Myuller (Müller), "Nature of the ionic conductivity of glass," Nature, 129(3257): 507 (1932).

4.   B. I. Markin and R. L. Myuller "A study of the conductivity of vitreous alkali metal borates," Zh. Fiz. Khim., 5(9):1262 (1934).

4a. B. I. Markin and R. L. Myuller (Müller), "Untersuchung der electrischan Leitfähigkeit glasartiger Alkaliborate," Acta Physicochim. URSS, 1(2):266 (1934).

4b. S. A. Shchukarev, R. L. Myuller, and B. I. Markin, "A study of the electrical conductivity of vitreous alkali metal borates" (Abstract Report), Protokoly of the Leningr. Nauchn. Issled. Inst. Khim. Obshchestva, 4:34 (1935).

5.   R. L. Myuller and B. I. Markin, "The nature of the electrical conductivity of low-alkali borate glasses," Zh. Fiz. Khim., 5(9):1272 (1934).

5a. R. L. Myuller (Müller) and B. I. Markin, "Zur Frage nach der Natur der elektrischen Leitfähigkeit der alkaliarmen Boratgläser," Acta Physicochim. URSS, 1(1):160 (1934).

6.   R. L. Myuller, "An experimental check on the theory of the conductivity of glasses," Zh. Fiz. Khim., 6(5):616 (1935).

6a. R. L. Myuller (Müller), Ein Versuch der theoretischen Erforschung der Leitfähigkeit von Gläsern., Acta Physicochim. URSS, 2(1):103 (1935).

6b. R. L. Myuller "An experimental check on the theory of the conductivity of glasses" (Abstract Report), Protokoly Zased. Leningr. Nauchn. Issled. Inst. Khim. Obshchestva, 4:34 (1935).

7.   B. Markin and R. L. Myuller, "A study of the electrical conductivity of vitreous barium borates," Zh. Fiz. Khim., 7(4):592 (1936).

7a. B. I. Markin and R. L. Myuller (Müller), "Untersuchung der elektrischen Leitfähigkeit der glasartigen Bariumborate," Acta Physicochim. URSS, 4(4):471 (1936).

8. R. L. Myuller, "The electrical conductivity of glasses," Report at the Jubilee Scientific Session for the 120th Anniversary of the Leningrad State University, Leningrad, (1939), p. 43.

9. R. L. Myuller, "Electrical conductivity of glasses," Uch. Zap. Leningr. Gos. Univ., No.54, p. 159 (1940).

10. R. L. Myuller, "The number of carriers in liquid and solid systems" (Abstract of report delivered at a meeting of the Leningrad section of the D. I. Mendeleev All-Union Chem. Soc,), Byull. Vses. Khim. Obshchestva im. D. I. Mendeleev, No. 7, p. 12 (1941).

11. R. L. Myuller, "Markin's work on the electrical conductivity of glasses," Zh. Tekhn. Fiz., 23(10):1874 (1953).

12. R. L. Myuller, "The nature of the electrical conductivity of glass," Zh. Éksper. Teor. Fiz., Vol. 27, No. 2(8), p. 264 (1954).

13. R. L. Myuller, Communication I, "The electrical conductivity of ionic-covalent materials," Introduction, Zh. Tekhn. Fiz., 25(2):236 (1955).

14. R. L. Myuller, Communication II, "Experimental and Theoretical expressions for the molar electrical conductivity of borosilicates," Zh. Tekhn. Fiz., 25(2):246 (1955).

15. R. L. Myuller, Communication III, "Polarization in an external field," Zh. Tekhn. Fiz., 25: 1556 (1955).

15a. The same (shortened version), Izv. Tomsk. Politekhn. Inst., 91:239 (1956).

16. R. L. Myuller, Communication IV, "Polarization and electrolytic phenomena," Zh. Tekhn. Zh. Tekhn. Fiz., 25(9):1567 (1955).

17. R. L. Myuller, Communication V, "The conductivity of borosilicates in the stable state," Zh. Tekhn. Fiz., 25(11):1868 (1955).

18. R. L. Myuller, Communication VI, "The conductivity of borosilicates in the labile state," Zh. Tekhn. Fiz., 25(14):2428 (1955).

19. R. L. Myuller, Communication VII, "The temperature dependence of the conductivity of crystalline materials," Zh. Tekhn. Fiz., 25(14):2440 (1955).

20. R. L. Myuller, Communication VIII, "The concentration dependence of the conductivity of borate and silicate glasses," Zh. Tekhn. Fiz., 26(12):2614 (1956).

21. R. L. Myuller, Communication IX, "The degree of dissociation and the cation mobility in glasses with one kind of ion," Fiz. Tverd. Tela, 2(6):1333 (1960).

22. R. L. Myuller, Communication X, "The electrical conductivity of glasses containing two kinds of alkali ions," Fiz. Tverd. Tela, 2(6):1339 (1960).

23. R. L. Myuller, Communication XI, "The degree of dissociation and the cation mobility in glass containing two kinds of ion," Fiz. Tverd. Tela, 2(6):1345 (1960).

24. R. L. Myuller, "The dependence of the conductivity of borosilicates on the metallic ion concentration," Izv. Tomsk. Politekhn. Inst., 91:353 (1956).

25. R. L. Myuller, "The relation between the electrical conductivity and the viscosity of glasses," Fiz. Tverd. Tela, 1(2):346 (1959).

26. R. L. Myuller, "The conductivity of complex glasses," in: Dielectric Physics, Proceeding of the Second All-Union Conference on the Vitreous State, Akad. Nauk SSSR, Moscow (1960), p. 438.

27. R. L. Myuller, "Cation mobilities and the degree of dissociation of polar groups as functions of the ionic and atomic composition of glasses," in: The Structure of Glass, Proceedings of the 3rd All-Union Conference, Akad. Nauk SSSR, Moscow—Leningrad (1960), p. 245 [English translation: The Structure of Glass, Vol. 2, Consultants Bureau, New York (1960), p. 215].

28. R. L. Myuller, "Electrical Properties of Glasses," Zh. Vses. Khim. Obshchestva im Mendeleeva, 8(2):197 (1963).

29. R. L. Myuller and A. A. Pronkin "Ionic conductivity of alkali aluminosilicate glasses," Zh. Prikl. Khim., 36(6):1192 (1963).

30. R. L. Myuller and A. A. Pronkin, "Nature of the conductivity of sodium aluminosilicate glasses," in: The Electrical Properties and Structure of Glass, Izd. "Khimiya," Moscow—Leningrad (1964), p. 51 [English translation: The Structure of Glass, Vol. 4, Consultants Bureau, New York (1965), p. 93].

31. R. L. Myuller, "Thermal ionization in glasses and mobility current carriers in them," ibid., p. 15 [p. 64].

32. R. L. Myuller and A. A. Pronkin "The poly-alkali effect in borosilicate glasses," in: The Chemistry of Solids, Izd. Leningr. Gos. Univ., (1965), p. 134.

33. V. S. Molchanov, R. L. Myuller and A. A. Pronkin, "The electrical conductivity of complex sodium—titanium—lead glasses," ibid., p. 146.

34. R. L. Myuller, and V. K. Leko, "The nature of electrical conductivity of non-alkali oxide glasses," ibid., p. 151.

35. R. L. Myuller and A. A. Pronkin, "Electrochemical data and the structures of some complex glasses," ibid., p. 173.

The Process of Glass Formation and the Electrical

Conductivity of Semiconducting Materials

1.   L. A. Baidakov, Z. U. Borisova, and R. L. Myuller, "Conductivity in the vitreous arsenic—selenium system," Zh. Prikl. Khim., 34(11):2446 (1961).

2.   R. L. Myuller, "The valence nature of the electrical conductivity of semiconducting glasses," Vestn. Leningr. Gos. Univ., Vol. 6, No. 22, Iss. 4, p. 86 (1961).

3.   R. L. Myuller, "Valence conductivity and structural chemical microhardness in vitreous semiconductors," in: Physics, Proc. of 20th Scientific Conference of the L. I. S. I., Izd. LISI, Leningrad (1962), p. 18.

4.   A. V. Danilov and R. L. Myuller, "Electrical Conductivity in the Vitreous System $AsSe_{1.5}$—Cu." Zh. Prikl. Khim., 35(9):2012 (1962).

4a. The same, in: Physics, Proc. 20th Scientific Conference of the L. I. S. I., Izd. LISI, Leningrad (1962), p. 21.

5.   R. L. Myuller, L. A. Baidakov, and Z. U. Borisova, "Electrical conductivity in the vitreous system As—Se—Ge," Vestn. Leningr. Gos. Univ., Vol. 17, No. 10, Iss. 2, p. 94 (1962).

5a. The same, in: Physics, Proc. of 20th Scientific Conference of the L. I. S. I. Izd. LISI, Leningrad (1962), p. 24.

6.   R. L. Myuller and T. P. Markova, "Electrical conductivity of the vitreous system, arsenic—selenium—thallium," Vestn. Leningr. Gos. Univ., Vol. 17, No. 4, Iss. 1, p. 75 (1962).

7.   R. L. Myuller, "The nature of the electrical conductivity of vitreous semiconductors," Zh. Prikl. Khim., 35(3):541 (1962).

8.   Z. U. Borisova, R. L. Myuller and Ch'in Ch'eng-Ts'ai, "The electrical conductivity of vitreous GeSe." Zh. Prikl. Khim., 35 (4):774 (1962).

9.   R. L. Myuller, L. A. Baidakov, and Z. U. Borisova, "An examination of the electrical conductivity in the vitreous system arsenic—sulfur," Vestn. Leningr. Gos. Univ., Vol. 17, No. 22, Iss. 4, p. 77 (1962).

10. R. L. Myller and E. V. Shkol'nikov, "A study of the crystallization of the glasses $As-Se_x-Ge_y$ by conductivity measurements," Vestn. Leningr. Gos. Univ., Vol. 17, No. 22, Iss. 4, p. 119 (1962).

11. R. L. Myuller, G. M. Orlova, V. N. Timofeeva, and G. I. Ternovaya, "The vitreous boundary in the system arsenic—sulfur—germanium," Vestn. Leningr. Gos. Univ., Vol. 17, No. 22, Iss. 4, p. 146 (1962).

12. E. V. Shkol'nikov, M. A. Rumsh, and R. L. Myuller, "An x-ray study of the crystallization of semiconducting glasses of the $As-Se_x-Ge_y$ type," Fiz. Tverd. Tela, 6(3):796 (1964).

13. R. L. Myuller, "Bond energy, ionization and the modulus of conductivity in vitreous semiconductors as functions of composition," Izv. Akad. Nauk SSSR, ser. fiz., 28(8):1279 (1964).

14. R. L. Myuller, M. El Mosli, and Z. U. Borisova, "The effect of heat treatment on the conductivity and microhardness of vitreous arsenic selenides," Vestn. Leningr. Gos. Univ., Vol. 19, No. 22, Iss. 4, p. 94 (1964).

15. R. L. Myuller, V. N. Timofeeva, and Z. U. Borisova, "A study of the electrical conductivity of the vitreous system arsenic—sulfur—germanium," Izd. Leningr. Gos. Univ., Leningrad (1965), p. 75.

Dissolution of Glasses

1. R. L. Myuller and Ts. V. Vainshtein, "Solution rates in the vitreous system $Me_2O-B_2O_3$ (Abstract Report), Protokoly Zased. Leningr. Otd. Nauchn. Issled. Inst. Khim. Obshchestva, 4:35 (1935).

2. Ts. V. Vainshtein and R. L. Myuller, "The solution rate of alkali borate glasses," Zh. Fiz. Khim., 7(3):364 (1936).

2a. R. L. Myuller (Müller) and Ts. V. Vainshtein (Weinstein), "Untersuchung der Lösunggeschwindigkeit von Akaliborgläsern," Acta Physicochim. URSS, 3(4):465 (1935).

3. R. L. Myuller, "Solution kinetics of alkali borate glasses," Zh. Fiz. Khim., 7(3):388 (1936).

3a. R. L. Myuller, Die Lösungskinetik der Alkaliborgläser, Acta Physicochim., URSS, 4(1):99 (1936).

4. R. L. Myuller and Ts. V. Vainshtein, "Etch figures in glasses" (Abstract Report), Protokoly Zased. Leningr. Otd. Nauchn. Issled. Inst. Khim. Obshchestva, 4:35 (1935).

4a. Ts. V. Vainshtein, B. I. Markin and R. L. Myuller, "Etch figures in glass," Zh. Fiz. Khim., 7, No. 3, 402 (1936).

4b. B. I. Markin, R. L. Myuller (Müller), and Ts. V. Vainshtein (Weinstein)," Zur Frage der 'Ätzfiguren' bei Gläsern," Acta Physicochim. URSS, 4(1):119 (1936).

5. R. L. Myuller, "A general expression for the solution rates of solution rates of solid materials," Zh. Fiz. Khim., 7(4):599 (1936).

5a. R. L. Myuller (Müller), Versuch der Auffindung einen gemeinsamen Ausdruckes für Lösungsgeschwindigkeit eines festen Körpers: Acta Physiocochim. URSS, 4(4):481 (1936).

6. R. L. Myuller, "Some essential features of physicochemical processes in heterophase nonmetallic materials," in: Scientific Proc. of the Kemerovo Mining Institute, No. 2, p. 160 (1956).

7. R. L. Myuller, R. Ts. Adzhemyan, and E. S. Shreiner, "The solubility of a covalent solid in a nonmobile liquid," Zh. Fiz. Khim., 36(8):1667 (1962).

8. R. L. Myuller and R. Ts. Adzhemyan "Solution kinetics of borax in aqueous solutions of electrolytes," Zh. Fiz. Khim., 36(9):1877 (1962).

9. R. L. Myuller and E. S. Shreiner, "Solution kinetics of borax in aqueous dioxane solutions," Zh. Fiz. Khim., 37(4):875 (1963).

Dissolution of Semiconductor Materials

1. R. L. Myuller, T. P. Markova, and S. M. Repinskii, "Solution kinetics of germanium in nitric acid," Vestn. Leningr. Gos. Univ., Vol. 14, No. 16, Iss. 3, p. 106 (1959).

2. R. L. Myuller, A. V. Danilov, T. P. Markova, V. N. Mel'nikov, A. B. Nikol'skii, and S. M. Repinskii, "Solution kinetics of germanium in acids and alkaline solutions of hydrogen peroxide," Vestn. Leningr. Gos. Univ., Vol. 15, No. 4, Iss. 1, p. 80 (1960).

3. R. L. Myuller and N. A. Baglai "The chemical kinetics of the solution of germanium in aqueous solutions of bromine and iodine," Vestn. Leningr. Gos. Univ., Vol. 15, No. 4, Iss. 1, p. 88 (1960).

4. R. L. Myuller, Z. U. Borisova, and N. I. Grebenshchikova, "Solution kinetics of arsenic selenide in an alkali medium," Zh. Prikl. Khim., 33:533 (1961).

5. R. L. Myuller, Z. U. Borisova, O. V. Il'inskaya, "Chemical micro-inhomogeneity of vitreous $AsS_{1.25}$ (solution kinetics of a complex solid)," Zh. Prikl. Khim. 33(3):690 (1961).

6. R. L. Myuller, G. M. Orlova, and Ts'ui Chin-hua, "Solution kinetics of indium antimonide in nitric acid," Zh. Obshch. Khim., 31(8):2457 (1961).

7. R. L. Myuller, G. M. Orlova, and Ts'ui Chin-hua, "Solution kinetics of indium antimonide in salt solutions of iron chloride and iodine chloride," Zh. Obshch. Khim., 31(8):2461 (1961).

The Thermal Capacity of Solids

1. R. L. Myuller "The thermal capacity of ionic-covalent solids," Zh. Fiz. Khim., 28(7):1193 (1954).

2. R. L. Myuller, "The theory of thermal capacity of vitreous heterodynamic structures," Zh. Fiz. Khim., 28(8):1521 (1954).

3. R. L. Myuller, "The chemistry of high-melting vitreous materials and thermal capacity data," Zh. Fiz. Khim., 28(10):1831 (1954).

4. R. L. Myuller, "The critical temperature region in silica from the thermal capacity data, and vitreous silicates," Zh. Fiz. Khim., 28(11):1954 (1954).

5. R. L. Myuller, "The critical temperature region in boric oxide from the thermal capacity data, and vitreous borates," Zh. Fiz. Khim., 28(12):2170 (1954).

6. R. L. Myuller, "Critical temperatures in low-melting glass and thermal capacity data," Zh. Fiz. Khim., 28(12):2189 (1954).

The Viscosity of Glasses

1. R. L. Myuller, "The nature of the activation energy and the experimental data for the fluidity of high melting vitreous materials," Zh. Prikl. Khim., 28(4):363 (1955).

2. R. L. Myuller, "The valence theory of viscosity and fluidity in the critical temperature region," Zh. Prikl. Khim., 28(10):1077 (1955).

3. R. L. Myuller, "The thermal capacity and viscosity of vitreous silicate materials," Tr. Tomsk. Gos. Univ., 145:33 (1957).

The Structure of Vitreous Materials

1. R. L. Myuller, "A physico-chemical analysis of vitreous systems using an electrical conductivity method," in: Proceedings of 1st. All-Union Conference on Physico-Chemical Analysis, Akad. Nauk SSSR, Leningrad (1933), p. 59.

2. R. L. Myuller, "The vitreous state" (Abstract report at a meeting of the Leningr. Sci. Res. Inst. Chem. Soc.), Byull. Vses. Khim. Obshchestva im. D. I. Mendeleeva, No. 6, p. 12 (1939).

3. R. L. Myuller, "The structure of solid glasses from the electrical conductivity data" Izv. Akad. Nauk SSSR, ser. fiz., 4(4):607 (1940).

4. R. L. Myuller, "The Vitreous State and the Electrochemistry of Glass" (Doctoral Dissertation), Leningr. Gos. Univ. (1940).

5. R. L. Myuller, "The structure of solid glasses from the electrical conductivity data" (Abstract Report at the conference on the vitreous state), Zh. Prikl. Khim., 13(3):479 (1940); Byull, Vses. Khim. Obshchestva im. D. I. Mendeleeva, No. 4, p. 47 (1940).

6. R. L. Myuller, "The vitreous state," Vestn. Znaniya, Nos. 7/8, p. 43 (1940).

7. R. L. Myuller, "The chemical structures of high-melting glasses," Zh. Fiz. Khim., 30(5):1146 (1956).

8. R. L. Myuller, "The vitreous state of materials," Steklo i Keramika, 13(4):11 (1956).

9. R. L. Myuller, "The Third All-Union Conference on the Vitreous State," Leningr. Gos. Univ., Vol. 16, No. 4, Iss. 2, p. 144 (1960).

10. R. L. Myuller, "The chemical characteristics of polymeric glass-forming substances, and the nature of glass formation," in: The Structure of Glass, Vol. 2, Proceedings of the 3rd. All-Union Conference, Izd. Akad. Nauk SSSR, Moscow—Leningrad (1960), p. 61. [English translation: The Structure of Glass, Vol. 2, Consultants Bureau, New York (1960), p. 50].

11. R. L. Myuller, "Cation mobilities and the degree of dissociation of polar groups as functions of the ionic and atomic composition of glass," in: The Structure of Glass, Vol. 2, Consultants Bureau, New York (1960) p. 215.

12. R. L. Myuller, "The structural chemical dependence of the refractive index of glasses," Zh. Prikl. Khim., 36(10):2154 (1963).

13. R. L. Myuller, "The structural chemical nature and the refractive index of glasses," Tr. Gos. Optich. Inst., 31(160):204 (1963).

14. R. L. Myuller, "The structure of glass and its crystallization," Priroda, 53(8):31 (1964).

15. R. L. Myuller, "Solid state chemistry and vitreous state," in: The Chemistry of the Solid State, Izd. Leningr. Gos. Univ. (1965), p. 9.

The Chemistry of the Platinum Metals

1. R. L. Myuller and E. Ya. Potepun-Afanas'eva, "The solution of palladium in nitric acid," Zh. Neorgan. Khim., 2(6):1306 (1957).

2. R. L. Myuller and V. M. Kostrikin, "An initial study of the chemical evaporation of ruthenium," Zh. Neorgan. Khim., 4(1):23 (1959).

3. A. A. Goryunov, R. L. Myuller, and L. K. Kapustina, "The rate of vaporization of ruthenium tetroxide from aqueous solutions by an air current," Leningr. Gos. Univ., Vol. 15, No. 10, Iss. 2, p. 104 (1960).

4. R. L. Myuller and A. B. Nikol'skii, "A method for the determination of the vapor pressure of ruthenium tetroxide vapor over its aqueous solutions using radioactive indicators," Radiokhimiya, 4(3):364 (1962).

Coal Extraction

1. R. L. Myuller, "The evolution of the energy base of human culture," Ugol' Strane, Nos. 60, 61, 62, 69, 72, Kemerovo (1954).

2. R. L. Myuller and V. S. Popov, "The kinetics of gas generation in connection with the problem of metamorphization of coal," Zh. Prikl. Khim., 30(2):271 (1957).

2a. The same, in: "Technical progress in the coal industry of the Kuzbas," Kemerovo (1957), p. 127.

3. R. L. Myuller, "The possible role of chemical processes in sudden outbursts of coal and gas," in: The Theory of Sudden Outbursts of Coal and Gas, Izd. Akad. Nauk SSSR, Moscow (1959), p. 156.

4. R. L. Myuller and V. S. Popov, "Methane formations in coals in connection with sudden outbursts of coal and gas in mines," in: Problems in Mining, No. 3, Gosgortekhizdat, Moscow (1960), p. 204.

Miscellaneous

1. R. L. Myuller, "A study of solutions by a streaming method," in: Contemporary Physical Chemical Methods of Chemical Analysis, No. 2, Khimteorizdat, Leningrad (1935), p. 108.

2. R. L. Myuller, A. V. Danilov, Yang Ying-Kuei, "Low temperature treatment of chemically deposited lead sulfide films," Zh. Prikl. Khim., 33(1):71 (1961).